AF576026

# Advances in Production, Properties and Applications of Sprouted Seeds

# Advances in Production, Properties and Applications of Sprouted Seeds

Editors

**Cristina Martínez-Villaluenga**
**Elena Peñas Pozo**

MDPI • Basel • Beijing • Wuhan • Barcelona • Belgrade • Manchester • Tokyo • Cluj • Tianjin

*Editors*

Cristina Martínez-Villaluenga
Spanish National Research Council (CSIC)
Spain

Elena Peñas Pozo
Spanish National Research Council (CSIC)
Spain

*Editorial Office*
MDPI
St. Alban-Anlage 66
4052 Basel, Switzerland

This is a reprint of articles from the Special Issue published online in the open access journal *Foods* (ISSN 2304-8158) (available at: https://www.mdpi.com/journal/foods/special_issues/sprouted_seeds).

For citation purposes, cite each article independently as indicated on the article page online and as indicated below:

LastName, A.A.; LastName, B.B.; LastName, C.C. Article Title. *Journal Name* **Year**, *Article Number*, Page Range.

**ISBN 978-3-03943-316-2 (Hbk)**
**ISBN 978-3-03943-317-9 (PDF)**

# Contents

**About the Editors** . . . . . . . . . . . . . . . . . . . . . . . . . . . . . . . . . . . . . . . . . . . . **vii**

**Elena Peñas and Cristina Martínez-Villaluenga**
Advances in Production, Properties and Applications of Sprouted Seeds
Reprinted from: *Foods* **2020**, *9*, 790, doi:10.3390/foods9060790 . . . . . . . . . . . . . . . . . . . . . . 1

**Carla S.Santos, Beatriz Silva, Luísa M.P.Valente, Sabine Gruber and Marta W.Vasconcelos**
The Effect of Sprouting in Lentil (*Lens culinaris*) Nutritional and Microbiological Profile
Reprinted from: *Foods* **2020**, *9*, 400, doi:10.3390/foods9040400 . . . . . . . . . . . . . . . . . . . . . . 5

**Miguel Rebollo-Hernanz, Yolanda Aguilera, Teresa Herrera, L. Tábata Cayuelas, Montserrat Dueñas, Pilar Rodríguez-Rodríguez, David Ramiro-Cortijo, Silvia M. Arribas and María A. Martín-Cabrejas**
Bioavailability of Melatonin from Lentil Sprouts and Its Role in the Plasmatic Antioxidant Status in Rats
Reprinted from: *Foods* **2020**, *9*, 330, doi:10.3390/foods9030330 . . . . . . . . . . . . . . . . . . . . . . **17**

**Daniel Rico, Elena Peñas, María del Carmen García, Cristina Martínez-Villaluenga, Dilip K. Rai, Rares I. Birsan, Juana Frias and Ana B. Martín-Diana**
Sprouted Barley Flour as a Nutritious and Functional Ingredient
Reprinted from: *Foods* **2020**, *9*, 296, doi:10.3390/foods9030296 . . . . . . . . . . . . . . . . . . . . . . **37**

**Gaetano Cardone, Paolo D'Incecco, Maria Cristina Casiraghi and Alessandra Marti**
Exploiting Milling By-Products in Bread-Making: The Case of Sprouted Wheat
Reprinted from: *Foods* **2020**, *9*, 260, doi:10.3390/foods9030260 . . . . . . . . . . . . . . . . . . . . . . **57**

**Margarita Damazo-Lima, Guadalupe Rosas-Pérez, Rosalía Reynoso-Camacho, Iza F. Pérez-Ramírez, Nuria Elizabeth Rocha-Guzmán, Ericka A. de los Ríos and Minerva Ramos-Gomez**
Chemopreventive Effect of the Germinated Oat and Its Phenolic-AVA Extract in Azoxymethane/Dextran Sulfate Sodium (AOM/DSS) Model of Colon Carcinogenesis in Mice
Reprinted from: *Foods* **2020**, *9*, 169, doi:10.3390/foods9020169 . . . . . . . . . . . . . . . . . . . . . . **75**

**Małgorzata Sikora, Michał Świeca, Monika Franczyk, Anna Jakubczyk, Justyna Bochnak and Urszula Złotek**
Biochemical Properties of Polyphenol Oxidases from Ready-to-Eat Lentil (*Lens culinaris* Medik.) Sprouts and Factors Affecting Their Activities: A Search for Potent Tools Limiting Enzymatic Browning
Reprinted from: *Foods* **2019**, *8*, 154, doi:10.3390/foods8050154 . . . . . . . . . . . . . . . . . . . . . . **97**

# About the Editors

**Cristina Martínez-Villaluenga** (PhD). B.S. in Biology from University Complutense of Madrid in 2001. PhD in Food Science by the University Autonoma of Madrid in 2006. She joined the Spanish Research Council (CSIC) in 2009. The long-term goal of Dr. Martínez's research program is to enhance the health of individuals by identifying and determining the benefits of plant food bioactive components with a special focus on bioactive peptides. Dr. Martínez's research on legumes, cereals, and pseudocereals has led to increased understanding of the anti-inflammatory, antihypertensive, antidiabetic, and other physiological properties of these foods. She is the author of 103 JCR articles and 9 book chapters with an h-index of 34 (WoS). Her results have been disseminated at 90 international and national conferences and over social media. In the last 10 years, she has supervised a total of 9 PhD theses, 5 Master theses, and more than 20 undergraduate students. She has participated in a total of 35 international and national R&D projects and contracts with the agrifood sector. She is a member of the editorial committees for 3 books and 3 journals.

**Elena Peñas Pozo** (PhD). B. S. in Biology from University Complutense of Madrid in 2000. PhD in Food and Technology by the University Autónoma of Madrid in 2006. She is a Tenured Scientist of the Institute of Food Science and Nutrition (ICTAN) belonging to the Spanish National Research Council (CSIC). Her research activities are aimed at optimizing technological bioprocesses for improving the safety, quality, and health-promoting properties of plant foods. The development and validation of novel foods from grains tailored to consumers with special nutritional needs is another of her research interests. She is the author of 80 JCR articles and 8 book chapters (h-index: 27). She has participated in 22 international and national R&D projects and contracts with the agrifood sector. Her results have been disseminated at 60 international and national conferences. In the last 10 years, she has supervised a total of 5 PhD theses, 5 Master theses, and more than 20 undergraduate students. She is a member of the editorial committees of 2 books and 1 journal.

*Editorial*

# Advances in Production, Properties and Applications of Sprouted Seeds

**Elena Peñas and Cristina Martínez-Villaluenga ***

Institute of Food Science, Technology and Nutrition (ICTAN-CSIC), Juan de la Cierva 3, 28006 Madrid, Spain; elenape@ictan.csic.es

* Correspondence: c.m.villaluenga@csic.es; Tel.: +34-91-258-76-01

Received: 4 June 2020; Accepted: 12 June 2020; Published: 16 June 2020

**Abstract:** Sprouted grains are widely appreciated food ingredients due to their improved, nutritional, functional, organoleptic and textural properties compared with non-germinated grains. In recent years, sprouting has been explored as a promising green food engineering strategy to improve the nutritional value of grains and the formation of secondary metabolites with potential application in the functional foods, nutraceutical, pharmaceutical and cosmetic markets. However, little attention has been paid to the impact of sprouting on the chemical composition, safety aspects, techno-functional and chemopreventive properties of sprouted seeds and their derived flours and by-products. The six articles included in this Special Issue provide insightful findings on the most recent advances regarding new applications of sprouted seeds or products derived thereof, evaluations of the nutritional value and phytochemical composition of sprouts during production or storage and explorations of their microbiological, bioactive and techno-functional properties.

**Keywords:** seed germination; nutritional value; phytochemicals; bioactivity; health; food safety; technological properties; food development; functional foods

---

Sprouted grains, usually designated as a seed with a visible radicle, have been used as food ingredients for many years, based on the general belief they provide significant nutritional, flavor, and textural benefits over non-germinated seed counterparts. In recent years, sprouting has been explored as a promising green food engineering strategy to improve the nutritional value of grains as well as to synthesize secondary metabolites with potential applications in the functional foods, nutraceutical, pharmaceutical and cosmetic markets. In this context, the industry has increasingly launched products containing or made of sprouted seeds. During seed sprouting, a multitude of changes occur, moving from molecular to macroscopic structures. Sprouting reactivates seed metabolism leading to the catabolism and degradation of macronutrients and antinutritional compounds and the biosynthesis of secondary metabolites with potential health benefits. These changes impact the nutritional value and health-promoting potential of the edible seeds. Many researchers around the world have proposed successful strategies such as elicitation to find the optimal environmental conditions during sprout growth able to promote the desired outcomes.

This Special Issue includes six outstanding papers describing examples of the most recent advances in new applications of sprouted seeds or products derived thereof, evaluations of the nutritional value, phytochemical composition and microbiological quality of sprouts during production or storage and explorations of their bioactive and techno-functional properties.

The Special Issue gathers a group of three papers exploring the biochemical composition, nutritional and bioactive properties as well as microbial safety of sprouts obtained from lentil seeds. In this frame, Santos et al. [1] conducted a study aimed at evaluating the protein and mineral profile of sprouts obtained from 12 lentils varieties. Protein, Zn, Mn, Ca and K contents were positively affected by sprouting in most of the lentil varieties studied, suggesting the potential of sprouting technology to improve the nutritional value of legumes. The authors also implemented a disinfection protocol combining SDS reagent and Amukine® application in order to ensure the microbial safety of lentils sprouts without affecting the germination rate and sprout length. Rebollo-Herran et al. [2] focused their research on the evaluation of the impact of lentil sprout intake on the plasmatic levels of melatonin and metabolically related compounds, total phenolic compounds and plasmatic antioxidant status compared to synthetic melatonin. The described results evidence that sprouting enhanced the levels of melatonin, decreasing the content of phenolic acids and flavan-3-ols in lentil. The administration of lentil sprouts to Sprague Dawley rats effectively increased the melatonin levels and antioxidant status in plasma, providing interesting insight on the beneficial effects of lentil sprouting on the attenuation of plasmatic oxidative stress mediated by melatonin. Sikora et al. [3] isolated and characterized polyphenol oxidase isoenzymes (I and II) from stored lentil sprouts and the mechanism of inhibition of these enzymes by antibrowning compounds and cations. The supplementation of sprouts with metal ions ($Zn^{2+}$, $Mn^{2+}$, $Fe^{3+}$) and/or inhibitors (ascorbic acid, citric acid) revealed to be an effective method to decrease the activity of polyphenol oxidase isoenzymes and to prevent enzymatic browning of lentil sprouts during storage and processing.

The Special Issue follows with a short series of articles describing the influence of sprouting on the chemical composition, bioactive and techno-functional properties of cereal flours and milling by-products. In this context, the study of Rico et al. [4] described the optimization of germination conditions to produce barley flour with a superior quality. The results reveal the applicability of sprouting in selected conditions as a valuable strategy to increase the nutritional value, phytochemical content and health benefits of barley. As a result, the authors obtained novel flours from sprouted barley enriched in vitamins, protein, gamma-aminobutyric acid (GABA) and antioxidant compounds. With the same objective, Cardone et al. [5] designed a study aimed at exploring the effect of sprouting on the chemical composition, enzymatic activities, techno-functional properties and bread-making performance of wheat bran, a milling by-product. The obtained results indicate that sprouting triggers positive nutritional changes in wheat bran by decreasing the antinutritional compounds and increasing the fiber content. Techno-functional properties such as water-holding capacity and gluten-aggregation kinetics were also improved as a consequence of the sprouting process, providing valuable characteristics to be used for bread formulation. The authors also demonstrate that breads formulated with wheat bran at a 20% of replacement level are enriched in fiber and show high-quality traits in terms of bread volume and crumb softness. This study evidences that sprouting offers interesting possibilities for the valorization of wheat milling by-products.

Another study conducted by Damazo-Lima et al. [6] reported the chemopreventive potential against colorectal cancer of sprouted oat and its phenolic-avenanthramide extract in a mouse model. Sprouted oat and its phenolic extract are valid chemopreventive ingredients when administered to animals since both reduce inflammation and tumor and adenocarcinoma incidence. Interestingly, sprouted oat exhibited a superior chemopreventive effect over its phenolic extracts, providing experimental evidence for a novel application of sprouted oat as a functional food for colon cancer prevention.

To conclude, the present Special Issue consists of six papers addressing recent advances on the nutritional, bioactive, techno-functional and safety aspects of sprouted grains and their derived flours and by-products, providing insightful information for both the food industry and consumers.

**Author Contributions:** Conceptualization, E.P. and C.M.-V.; validation, E.P. and C.M.-V.; resources, E.P. and C.M.-V.; writing—original draft preparation, E.P. and C.M.-V.; writing—review and editing, E.P. and C.M.-V.; visualization, E.P. and C.M.-V.; supervision, E.P. and C.M.-V.; funding acquisition, E.P. and C.M.-V. All authors have read and agreed to the published version of the manuscript.

**Funding:** This editorial letter has been prepared within the frame of grant number AGL2017-83718-R funded by FEDER/Ministry of Science, Innovation and Universities-State Agency of Research (AEI/Spain and FEDER/UE).

**Conflicts of Interest:** The authors declare no conflicts of interest.

## References


1. Santos, C.S.; Silva, B.; Valente, L.M.P.; Gruber, S.; Vasconcelos, M.W. The Effect of Sprouting in Lentil (*Lens culinaris*) Nutritional and Microbiological Profile. *Foods* **2020**, *9*, 400. [CrossRef] [PubMed]
2. Rebollo-Hernanz, M.; Aguilera, Y. Bioavailability of Melatonin from Lentil Sprouts and Its Role in the Plasmatic Antioxidant Status in Rats. *Foods* **2020**, *9*, 330. [CrossRef] [PubMed]
3. Sikora, M.; Swieca, M.; Franczyk, M.; Jakubczyk, A.; Bochnak, J.; Zlotek, U. Biochemical Properties of Polyphenol Oxidases from Ready-to-Eat Lentil (*Lens Culinaris* Medik.) Sprouts and Factors Affecting Their Activities: A Search for Potent Tools Limiting Enzymatic Browning. *Foods* **2020**, *8*, 154. [CrossRef]
4. Rico, D.; Peñas, E.; García, M.C.; Martínez-Villaluenga, C.; Rai, D.K.; Birsan, R.I.; Frias, J.; Martín-Diana, A.B. Sprouted Barley Flour as a Nutritious and Functional Ingredient. *Foods* **2020**, *9*, 296. [CrossRef] [PubMed]
5. Cardone, G.; D'Incecco, P.; Casiraghi, M.C.; Marti, A. Exploiting Milling By-Products in Bread-Making: The Case of Sprouted Wheat. *Foods* **2020**, *9*, 260. [CrossRef] [PubMed]
6. Damazo-Lima, M.; Rosas-Pérez, G.; Reynoso-Camacho, R.; Pérez-Ramírez, I.F.; Rocha-Guzmán, N.E.; de los Ríos, E.A.; Ramos-Gomez, M. Chemopreventive Effect of the Germinated Oat and Its Phenolic-AVA Extract in Azoxymethane/Dextran Sulfate Sodium (AOM/DSS) Model of Colon Carcinogenesis in Mice. *Foods* **2020**, *9*, 169. [CrossRef] [PubMed]

*Article*

# The Effect of Sprouting in Lentil (*Lens culinaris*) Nutritional and Microbiological Profile

**Carla S.Santos [1,*], Beatriz Silva [1], Luísa M.P.Valente [2,3], Sabine Gruber [4] and Marta W.Vasconcelos [1]**

1 Universidade Católica Portuguesa, CBQF—Centro de Biotecnologia e Química Fina—Laboratório Associado, Escola Superior de Biotecnologia, Rua Diogo Botelho 1327, 4169-005 Porto, Portugal; 19.beatriz.98@gmail.com (B.S.); mvasconcelos@porto.ucp.pt (M.W.V.)

2 CIIMAR – Centro Interdisciplinar de Investigação Marinha e Ambiental, Universidade do Porto, Avenida General Norton de Matos, 4450-208 Matosinhos, Portugal; lvalente@icbas.up.pt

3 ICBAS, Instituto de Ciências Biomédicas de Abel Salazar, Universidade do Porto, Rua de Jorge Viterbo Ferreira, 228, 4050-313 Porto, Portugal

4 Universität Hohenheim, Institut für Kulturpflanzenwissenschaften, Fg. 340a Allgemeiner Pflanzenbau, Fruwirthstr. 23, 70599 Stuttgart, Germany; Sabine.Gruber@uni-hohenheim.de

* Correspondence: cssantos@porto.ucp.pt

Received: 27 February 2020; Accepted: 20 March 2020; Published: 1 April 2020

**Abstract:** Biological and vegetarian raw food products, in particular based on legume sprouts, are an increasing food trend, due to their improved nutritional value when compared to seeds. Herein, protein and mineral profiles were studied in 12 lentil varieties, with varieties Du Puy, Kleine Schwarze, Rosana, Flora, Große Rote and Kleine Späths II demonstrating the highest protein percentages. After sprouting, protein percentages increased significantly in 10 of the 12 varieties, with the highest increases ranging between 20–23% in Dunkelgrün Marmorierte, Du Puy, Große Rote and Kleine Späths II varieties. While Fe concentration was significantly decreased in three varieties (Samos, Große Rote and Kleine Späths II), Zn and Mn were positively impacted by sprouting ($p \leq 0.05$). Magnesium concentration was not affected by sprouting, while Ca and K had percentage increases between 41% and 58%, and 28% and 30%, respectively, in the best performing varieties (Kleine Schwarze, Dunkelgrün Marmorierte, Samos and Rosana). Regardless of the associated nutritional benefits, issues pertaining to sprouts microbiological safety must be ensured. The best results for the disinfection protocols were obtained when combining the seed treatment with SDS reagent followed by an Amukine application on the sprouts, which did not affect germination rates or sprout length. The increasing levels of sprout consumption throughout the world require efficient implementation of safety measures, as well as a knowledge based selection for the nutritional quality of the seeds.

**Keywords:** food safety; legumes; microbial contamination; protein; mineral

## 1. Introduction

The consumption of legume seeds and germinated sprouts is increasing, being considered "functional foods" due to their increased nutrient availability and bioactive compounds [1,2]. Furthermore, sprouting the seeds has beneficial effects over seed quality, namely increasing digestibility and reducing the content of resistant starch and anti-nutritive compounds [3].

Lentil (*Lens culinaris* L.), in addition to having high protein content, low caloric value and high levels of essential nutrients such as folate, vitamin C and fibre [4], has as an advantage, when compared to other legumes and cereals, of very low phytic acid concentration [5], as well as high total phenolic levels [6]. Hence, this legume is a good source of amino acids, nutrients and high-quality protein [7].

Additionally, studies link lentil consumption with decreased body weight and body fat [8] and antihypertensive function [9].

Although legumes are considered a healthy option, over the years there have been pathogenic outbreaks associated with the consumption of seeds and raw sprouts [10–12]. Many of these outbreaks occur following seed and seed sprout consumption contaminated with *Escherichia coli*, *Salmonella* spp. and sometimes *Listeria monocytogenes* [11,13]. For example, in 2011 an *E. coli* outbreak in Germany was originated from contaminated seeds from Egypt. Another outbreak that occurred in Canada in 2005, linked to the consumption of contaminated sprouts, resulted in 600 people infected [14]. These cases demonstrate that seeds and sprouts can be easily contaminated, as also shown in a recent study where *L. innocua*, *Salmonella* spp. and coliform bacteria were found in microgreens and seed sprouts in Latvia [15]. Moreover, consumer demand for food that marketed as natural has resulted in the reduction of pesticides and other compounds, making it more likely to encounter infected seeds.

It is therefore necessary to ensure the safety of these products by developing methods that the consumer can use in the washing and disinfection of these foods at home, because they are usually consumed raw or lightly cooked [15,16]. Common household disinfection methods include hot water, acetic acid and Amukine applications, whereas sodium dodecyl sulphate (SDS), sodium hypochlorite and ethanol are common reagents used in laboratory applications for material disinfection [17].

The main objectives of this study were to comprehend the impact of sprouting on the nutritional profile of a collection of 12 lentil varieties and to understand the efficacy of different disinfection methods at eliminating *E. coli* and *Salmonella* spp. in both seeds and seed sprouts, without compromising germination percentage and sprout length.

## 2. Materials and Methods

### 2.1. Plant material and Seed Germination

A field trial was set up at the research station Kleinhohenheim that is located in south-west Germany near the city of Stuttgart (48°44′N, 9°11′O; 435 m a.s.l.). The research station has been managed organically since 1993. The climate is temperate (Cfb according to Köppen-Geiger classification [18]) and, during the experiment, which occurred from 6th April 2016 to 28th July 2016, the mean temperature was 14.9 °C and the sum of precipitation was 264 mm. At the summer solstice, there were 16 h 7 min between sunrise and sunset at the location. The soil type of the field was a loess-born silty loam with good drainage, with a pH of 6.3, total mineral N of 48 kg/ha and 9 mg/100 g of $P_2O_5$, 18 mg/100 g of $K_2O$ and 12 mg/100 g of Mg. In April 2016, the area was prepared for seeding by a rotor harrow and 12 lentil germplasms were sown by hand in double rows of 1 m length with a row spacing of 15 cm, and a target plant density of 240 lentil plants $m^{-2}$. To avoid lodging, a fence of mesh wire was installed for all plots where the lentils fixed themselves by their tendrils. During the experimental period, the plants were healthy and did not present any signs of diseases or infestations. Twelve different lentil varieties were harvested and stored at 4 °C until further studies: 1—Dunkelgrün Marmorierte; 2—Du Puy; 3—Thessalia; 4—Dimitra; 5—Samos; 6—Kleine Schwarze; 7–Rosana; 8—Flora; 9—Santa; 10—Große Rote; 11—Kleine Rote; 12—Späths Alblinse II 'Die Kleine'.

Seeds were germinated according to the protocol used by Shanmugam et al. [19]. Briefly, 50 seeds were placed in a beaker, covered with 70 % ethanol, and left for five minutes with agitation. The ethanol was discarded and a solution of 1.2% sodium hypochlorite and 0.02% SDS was added to cover the seeds and was left for 15 minutes with agitation. The solution was discarded and seeds were rinsed five times with deionized water. Afterwards, the seeds were germinated in Petri dishes with two bottom layers of paper filter moistened with deionized water, in the dark, at room temperature. At the end of five days, the lentil sprouts were stored in liquid nitrogen and then lyophilized for the nutritional analyses. This experiment was conducted in triplicate for all 12 varieties. To calculate the percentage of germination the following formula was applied:

$$\% \, germination = \frac{sprouts}{total \, seeds} * 100 \quad (1)$$

### 2.2. Nutritional Analysis

Samples of seeds and seed sprouts of the 12 lentil varieties ($n = 3$) were analysed for minerals and protein. Mineral analysis determination was performed as described by Santos et al. [20]. The minerals analysed were iron (Fe), zinc (Zn), manganese (Mn), magnesium (Mg), calcium (Ca) and potassium (K). Briefly, 200 mg of the dried seed or seed sprout material was mixed with 6 mL of 65% $HNO_3$ and 1 mL of 30% $H_2O_2$ in a Teflon reaction vessel and heated in a SpeedwaveTM MWS-3+ (Berghof, Germany) microwave system. Digestion procedure was conducted in five steps, consisting of different temperature and time sets: 130 °C/10 min, 160 °C/15 min, 170 °C/12 min, 100 °C/7 min, and 100 °C/3 min. The resulting clear solutions of the digestion procedure were then brought to 50 mL with ultrapure water for further analysis. Mineral concentration determination was performed using the ICP-OES Optima 7000 DV (PerkinElmer, Waltham, MA, USA) with radial configuration.

Seeds and seed sprouts were analysed for crude protein concentration (N × 5.28) using a Leco nitrogen analyzer (Model FP-528, Leco Corporation, St. Joseph, MO, USA).

### 2.3. Preparation of Inocula and Seed Inoculation

To ensure seed contamination for optimizing seed disinfection methods, seeds of the *Rosana* variety were inoculated according to the protocol used in [4]. In short, two solutions of 200 mL of Buffered Peptone Water nutrient medium (BPW) were prepared with 2 mL of *E. coli* and 2 mL of *Salmonella* spp. inocula. Using these solutions, 60 g were inoculated with *E. coli* $1.0 \times 10^{-8}$ UFC/ml and another 60 g were inoculated with *Salmonella* spp. $1.0 \times 10^{-8}$ UFC/ml. Seed samples were incubated for five minutes with gentle agitation. An additional 60 g of seeds was incubated with no bacteria inocula as control. After decanting the supernatant, the seeds were placed on a tray lined with filter paper and dried in a biosafety cabinet at room temperature (approximately 20 °C) for eight hours to determine the seed bacterial load.

### 2.4. Seed Contamination Evaluation

Seeds—inoculated and control—were placed in different sterile stomacher bags with buffered peptone water (BPW) until making a 1:10 dilution. Afterwards, the seed samples went to the stomacher in cycles of approximately 10 seconds at a time until a total of approximately one minute. The bacterial load on untreated and treated *Rosana* seeds was determined by the plate count method. McConkey Agar was used to plate *E. coli* and Rapid Agar *Salmonella* was used to plate *Salmonella* spp. Plates were incubated at 37 °C for 24 h. The results obtained were then converted to UFC/mL by using the following formula:

$$UFC/mL = \frac{Number \, of \, colonies * dilution \, factor}{volume \, of \, culture \, plate} \quad (2)$$

### 2.5. Seed and Seed Sprout Disinfection Methods

Two methods of seed disinfection were compared: (1) 70% ethanol for five minutes followed by 15 minutes of a solution of 1.2% sodium hypochlorite and 0.02% SDS; and (2) hot water treatment [4], which consisted of placing the seeds in deionized water at 80 °C for 90 s, followed by drying the seeds on a sterile paper filter.

For the first method, 20 g of inoculated seeds with *E. coli* and *Salmonella* spp. were disinfected and then the bacterial load on the seed was determined by the plate count method. The same procedure was followed when testing the hot water method. Afterwards, the seeds were germinated following germination protocol described previously. The resulting sprouts were measured and went through two more disinfection protocols: cleansing with water; and treatment with Amukine, following manufacturer instructions: 15 min, 50 mL for 2.5 L of water. After these two procedures, the microbial charge on the seeds and seed sprouts was determined by the plate count method.

### 2.6. Statistical Analysis

All data were analysed with GraphPad Prism version 6.00 for Mac OS X (GraphPad Software, La Jolla, CA, USA [21]) using Tukey's test.

## 3. Results and Discussion

### 3.1. Germination Efficiency

Germination efficiency and nutritional analyses were performed in the 12 lentil varieties to select the best performing seed variety for the microbiology study.

Germination is a bioprocess in which dry pulse seeds move from a dormant state to a metabolically and cellularly active state [22]. In Figure 1, the germination rates of each lentil variety were assessed and Du Puy (88%), Rosana (93%), Kleine Rote (88%) and Kleine Späths II (89%) were the highest performing varieties.

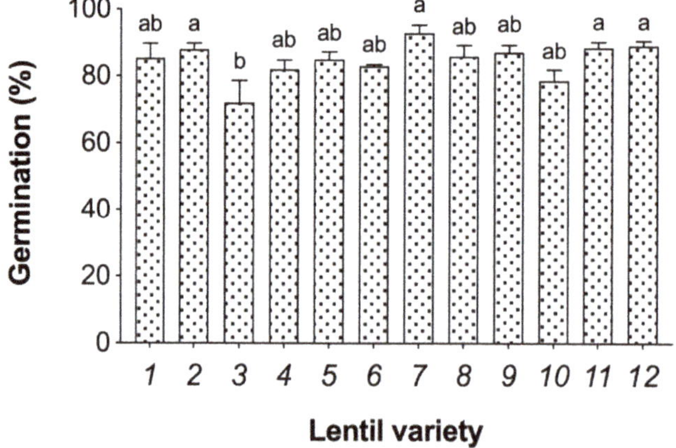


**Figure 1.** Germination rates of the 12 lentil varieties: 1—Dunkelgrün Marmorierte; 2—Du Puy; 3—Thessalia; 4—Dimitra; 5—Samos; 6—Kleine Schwarze; 7—Rosana; 8—Flora; 9—Santa; 10—Große Rote; 11—Kleine Rote; 12—Kleine Späths II. The bars represent means ± SE ($n$ = 3); different letters indicate significant differences ($p \leq 0.05$) by Tukey's Test.

### 3.2. Nutritional Analysis

In terms of physicochemical properties, unlike other pulses, such as chickpea, fava bean or pea, lentil germplasms are relatively stable and their pasting properties or hydration capacity vary little across different accessions/varieties [23,24]. This is relevant since these characteristics are significantly correlated to the nutritional quality of the seeds, herein analysed.

In the case of protein concentration, in the present study, a significant variation was found between the lentil varieties ($p \leq 0.05$), and this intraspecific variation was also observed in a different group of 12 lentil varieties analysed in a different study [25]. The mean protein concentration amongst seeds was 24.4% and the highest measured varieties (with values above the mean) were Du Puy, Kleine Schwarze, Rosana, Flora, Große Rote, Kleine Späths II.

Furthermore, after seed germination, the protein concentration was highly increased in all lentil varieties (Figure 2), and mean protein concentration in seed sprouts was 29%. For varieties with significant variation between protein levels in seeds and seed sprouts, the ones with the highest percentage increases were Dunkelgrün Marmorierte (23%), Du Puy (20%), Große Rote (22%) and Kleine Späths II (20%). Comparative studies using different pulses also showed that lentil has the highest protein content and that total protein values increase after germination [22,25]. In the present study, protein was measured as total N content, which has been reported to remain unaltered by germination [26]. Thus, the higher values of protein here reported are promising but should be confirmed in future studies.

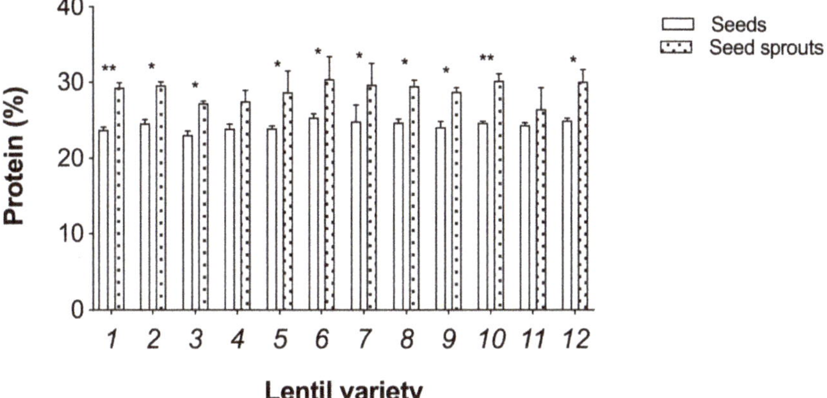


**Figure 2.** Protein concentration (%) of the seeds and seed sprouts of 12 lentil varieties: 1—Dunkelgrün Marmorierte; 2—Du Puy; 3—Thessalia; 4—Dimitra; 5—Samos; 6—Kleine Schwarze; 7—Rosana; 8—Flora; 9—Santa; 10—Große Rote; 11—Kleine Rote; 12—Kleine Späths II. Bars represent means ± SE ($n = 3$). * and ** indicate significant differences between seeds and seed sprouts at $p \leq 0.05$ and $p \leq 0.01$ respectively, by ANOVA using Tukey's Test.

Regarding seed and seed sprout mineral concentrations, six nutrients were selected for the present analysis. The selected micronutrients were Fe, Zn and Mn and the macronutrients Mg, Ca and K (Table 1).

As observed in other studies [27], Fe concentration values were on average ~50 µg/g (Table 1). The lentil varieties with highest Fe concentration (both in seeds and seed sprouts) were Thessalia, Dimitra, Rosana, Flora and Kleine Rote. Germination process leads to a significant decrease in Fe concentration of Samos (36%), Große Rote (11%) and Kleine Späths II (14%) varieties. This effect of germination in Fe concentration has been reported in previous studies using different legumes seeds, namely, soybean and kidney bean [28,29]. These studies showed that the seed Fe concentration decrease was counterbalanced by a major improvement in the availability of Fe. The concentration of the micronutrients Zn and Mn, on the contrary, were shown to be positively impacted by germination, as obtained here (Table 1).

**Table 1.** Iron (Fe), zinc (Zn), manganese (Mn), magnesium (Mg), calcium (Ca) and potassium (K) concentration (μg/g) of the seeds and seed sprouts of 12 lentil varieties and correspondent *P* values of the differences between them by ANOVA using Tukey's Test.

| Lentil Varieties | | Nutrient Concentration (μg/g) | | | | | | | | | | | |
|---|---|---|---|---|---|---|---|---|---|---|---|---|---|
| | | **Fe** | ***p* Value** | **Zn** | ***p* Value** | **Mn** | ***p* Value** | **Mg** | ***p* Value** | **Ca** | ***p* Value** | **K** | ***p* Value** |
| Dunkelgrün Marmorierte | Seeds | 38 ± 0.3 | n.s. | 31 ± 0.07 | <0.0001 | 11 ± 0.4 | n.s. | 10005 ± 23 | n.s. | 681 ± 23 | 0.0013 | 9994 ± 308 | n.s. |
| | Sprouts | 37 ± 0.1 | | 40 ± 0.9 | | 13 ± 1.8 | | 952 ± 11 | | 1079 ± 73 | | 12,130 ± 372 | |
| Du Puy | Seeds | 37 ± 0.9 | n.s. | 30 ± 0.04 | <0.0001 | 11 ± 0.07 | <0.0001 | 991 ± 22 | n.s. | 619 ± 36 | 0.0109 | 9730 ± 303 | n.s. |
| | Sprouts | 39 ± 0.3 | | 36 ± 0.2 | | 23 ±4.1 | | 891 ± 18 | | 964 ± 36 | | 10,023 ± 187 | |
| Thessalia | Seeds | 50 ± 0.2 | n.s. | 49 ± 0.1 | n.s. | 17 ± 0.2 | n.s. | 1148 ± 48 | n.s. | 933 ± 40 | n.s. | 10,769 ± 889 | n.s. |
| | Sprouts | 53 ± 0.5 | | 48 ± 0.4 | | 18 ± 0.5 | | 1247 ± 104 | | 774 ± 66 | | 10,830 ± 394 | |
| Dimitra | Seeds | 59 ± 0.3 | n.s. | 42 ± 0.2 | n.s. | 13 ± 0.3 | n.s. | 1155 ± 101 | n.s. | 1040 ± 38 | n.s. | 10,296 ± 641 | n.s. |
| | Sprouts | 63 ± 0.8 | | 44 ± 0.6 | | 14 ± 0.8 | | 928 ± 25 | | 866 ± 35 | | 10,830 ± 705 | |
| Samos | Seeds | 53 ± 0.2 | <0.0001 | 39 ± 0.3 | <0.0001 | 13 ± 0.2 | n.s. | 1067 ± 36 | n.s. | 999 ± 64 | n.s. | 9623 ± 303 | n.s. |
| | Sprouts | 34 ± 1.6 | | 48 ± 0.8 | | 14 ± 0.6 | | 1039 ± 45 | | 1138 ± 52 | | 12,299 ± 383 | |
| Kleine Schwarze | Seeds | 47 ± 0.3 | n.s. | 44 ± 0.6 | <0.0001 | 14 ± 0.3 | n.s. | 1201 ± 83 | n.s. | 754 ± 70 | 0.0350 | 11,050 ± 375 | n.s. |
| | Sprouts | 50 ± 1.7 | | 51 ± 0.3 | | 10 ± 0.4 | | 1043 ± 25 | | 1067 ± 68 | | 10,292 ± 169 | |
| Rosana | Seeds | 68 ± 1.5 | n.s. | 35 ± 0.5 | n.s. | 12 ± 0.5 | 0.0082 | 947 ± 70 | n.s. | 775 ± 62 | 0.0026 | 10,149 ± 68 | 0.0419 |
| | Sprouts | 64 ± 3.1 | | 38 ± 0.6 | | 20 ± 0.8 | | 1017 ± 52 | | 1156 ± 113 | | 13,154 ± 1043 | |
| Flora | Seeds | 51 ± 0.3 | n.s. | 37 ± 0.4 | n.s. | 12 ± 0.7 | n.s. | 1107 ± 73 | n.s. | 851 ± 62 | n.s. | 9414 ± 807 | n.s. |
| | Sprouts | 52 ± 0.4 | | 39 ± 0.09 | | 12 ± 0.5 | | 1010 ± 55 | | 759 ± 55 | | 11,350 ± 491 | |
| Santa | Seeds | 43 ± 0.4 | n.s. | 36 ± 1.5 | n.s. | 13 ± 0.9 | n.s. | 1168 ± 67 | n.s. | 766 ± 4 | n.s. | 10,905 ± 668 | n.s. |
| | Sprouts | 43 ± 0.3 | | 36 ± 0.2 | | 16 ± 0.6 | | 1044 ± 72 | | 668 ± 30 | | 10,128 ± 192 | |
| Große Rote | Seeds | 51 ± 0.2 | 0.0374 | 38 ± 1.4 | 0.0186 | 15 ± 2 | 0.0079 | 1006 ± 47 | n.s. | 864 ± 52 | n.s. | 10,905 ± 668 | n.s. |
| | Sprouts | 45 ± 0.5 | | 42 ± 1.1 | | 23 ± 0.6 | | 1061 ± 27 | | 980 ± 35 | | 10,128 ± 192 | |
| Kleine Rote | Seeds | 66 ± 0.2 | n.s. | 43 ± 0.3 | 0.0195 | 17 ± 0.6 | 0.0167 | 1084 ± 17 | n.s. | 1062±70 | n.s. | 10,423 ± 849 | n.s. |
| | Sprouts | 63 ± 2.2 | | 47 ± 1.1 | | 25 ± 1.1 | | 1048 ± 28 | | 1146 ± 62 | | 10,216 ± 308 | |
| Kleine Späths II | Seeds | 44 ± 0.2 | 0.0152 | 38 ± 0.2 | <0.0001 | 11 ± 1.2 | n.s. | 1224 ± 121 | n.s. | 623 ± 52 | n.s. | 10,121 ± 211 | n.s. |
| | Sprouts | 38 ± 0.9 | | 47 ± 0.1 | | 16 ± 1.5 | | 1075 ± 54 | | 803 ± 21 | | 10,473 ± 425 | |

n.s. not significant.

More specifically, of the 12 varieties under analysis, only five did not register a significant increase in Zn concentration after sprouting. In general, the varieties with higher Zn concentration were Thessalia, Dimitra, Kleine Schwarze and Kleine Rote. Higher Zn concentration increases ($p \leq 0.0001$) were found in Dunkelgrün marmorierte (29%), Du Puy (21%), Samos (22%) and Kleine Späths II (24%) varieties. In regards to Mn concentrations (Table 1), Thessalia, Große Rote and Kleine Rote demonstrated the highest both in seeds and seed sprouts; and the varieties that presented a significant increase after germination were Du Puy (108%), Rosana (64%), Große Rote (52%) and Kleine Rote (42%). Legume germination is associated with a drastic reduction in phytate content, which in seeds bind with minerals, forming insoluble complexes, making them unavailable [29,30].

Amongst the analysed macronutrients, Mg concentration was not affected by germination (Table 1), similarly to what was found in a study using soybean seeds [29]. Calcium concentration, on the other hand, was reported to increase approximately 55% in legume seeds after germination [30,31], as well as its bioavailability [28]. In the present study, the varieties with the highest Ca concentration (Samos and Kleine Rote) did not show variations after germination. The varieties in which Ca concentration was significantly increased were Dunkelgrün Marmorierte (58%), Du Puy (56%), Kleine Schwarze (41%) and Rosana (49%).

In terms of K concentrations, it was reported that soybean seed sprouts present a concentration five-times higher when compared to dry seeds [29]. In the present study (Table 1), the varieties with the highest K concentration were Thessalia, Kleine Schwarze, Santa and Große Rote. Seed K concentration has been identified as a possible marker for germination capacity due to its role in initiating the imbibition of water and facilitating the associated physiological processes [32,33]. Here, the varieties with the highest K concentration (Table 1) are not amongst the ones with significantly higher germination rates (Figure 1); however, significant differences amongst varieties were few. Only Rosana variety presented significant increases (30%) in K concentration after germination (Table 1).

As all 12 varieties were grown in the same field conditions, the differences detected in this study are mainly genotypic. Interestingly, Dunkelgrün Marmorierte and Du Puy varieties, that share the same genetic background, showed no significant differences for the analysed factors, both in the seeds and seed sprouts. Knowledge on the correlation between genotypic variation and nutritional traits can contribute to future breeding programs as well as for a targeted selection of the most appropriate varieties for human consumption.

Between the four varieties with the highest germination rate—Rosana, Kleine Späths II, Kleine Rote and Du Puy—and based on their percentage of protein and potential for mineral increase after germination, the following studies proceeded with the Rosana variety.

### 3.3. Microbial Counting and Disinfecting Methods

Firstly, the impact of the seed disinfection treatments on lentil germination percentage was tested (Table 2). It was found that in general, the germination efficiencies for all treatments and controls were high, with most seeds germinating at over 90%. In general, there was no negative impact of disinfection on germination efficiency, albeit the hot water treatment for *Salmonella* inoculated seeds seems to have lowered the germination values by about 32.6% when compared to the non-disinfected non-inoculated seeds and non-disinfected and inoculated seeds.

In order to test the different disinfection treatments, the seeds were artificially inoculated with *Salmonella* spp. and *E. coli*. As shown in Table 3, disinfection with only hot water was efficient for reducing *E. coli* (from $2.7 \times 10^8$ to $2.7 \times 10^7$ UFC/mL), as was SDS treatment for *Salmonella* spp. (from $1.1\times 10^8$ UFC/mL to $\leq 1.0 \times 10^8$ UFC/mL). Regarding sprouts, the reduction was more accentuated when the combination of Amukine and SDS was applied, reducing 2 logs (corresponding to a 99% reduction of bacterial load) for *E. coli* when compared to sprouts inoculated with no disinfection and 1 log logs (corresponding to a 90% reduction of bacterial load) for *Salmonella* spp.

**Table 2.** Germination percentage after disinfection treatments in lentil seeds (variety Rosana) inoculated with *E. coli*, *Salmonella* spp. and control.

| Treatments | % Germination | | |
|---|---|---|---|
| | *E. coli* | *Salmonella* spp. | Control |
| No disinfection/without inoculation | - | - | 98.8 |
| No disinfection/after inoculation | 99.3 | 96.9 | - |
| SDS disinfection | 96.3 | 96.0 | 97.3 |
| Water (80 °C) disinfection | 95.5 | 65.3 | 82.3 |

**Table 3.** Microbial counting after disinfection treatments (UFC/mL) of lentil seeds and sprouts.

| Treatments | Seeds | | Sprouts | |
|---|---|---|---|---|
| | *E. coli* | *Salmonella* | *E. coli* | *Salmonella* |
| **No disinfection/without inoculation** | $\leq 1.0 \times 10^8$ | $\leq 1.0 \times 10^8$ | | |
| **No disinfection/with inoculation** | $2.7 \times 10^8$ | $1.1 \times 10^8$ | $1.8 \times 10^8$ | $1.4 \times 10^8$ |
| **SDS disinfection** | $1.1 \times 10^8$ | $\leq 1.0 \times 10^8$ | $11.4 \times 10^8$ | $2.7 \times 10^8$ |
| **Water (80 °C) disinfection** | $2.7 \times 10^7$ | $>3.0 \times 10^8$ | $2.1 \times 10^8$ | $1.2 \times 10^8$ |
| **Water (80 °C) + $H_2O$ rinse** | | | $9.0 \times 10^7$ | $9.4 \times 10^7$ |
| **Water (80 °C) + Amukine** | | | $9.8 \times 10^7$ | $8.0 \times 10^7$ |
| **SDS + $H_2O$ rinse** | | | $6.0 \times 10^7$ | $9.8 \times 10^8$ |
| **SDS + Amukine** | | | $3.9 \times 10^6$ | $5.6 \times 10^7$ |

In 1999, the U.S. Food and Drug Administration recommended the utilization of 20,000 ppm of calcium hypochlorite for seed disinfection [34] and this is the method commonly used by sprout manufacturers. However, this treatment was considered potentially hazardous to the environment and industrial workers, and new efficient strategies are needed to improve sprout safety [16].

The products selected for this experiment are of easy access and are usually utilized for disinfection since they have chemical constituents, such as hypochlorite (found in common household bleach and Amukine) and ammonia, that act by denaturing bacterial proteins, similar to what happens when exposed to high temperatures resulting in bacterial death [35], or when exposed to ethanol and sodium hypochlorite. Here, some combinations of these treatments were tested to mimic and improve the disinfections done in a domestic environment.

As shown in Table 1, in general, the germination rate was not affected by the treatments or the inoculation with the pathogens, as they presented similar germination rates. However, seeds inoculated with *Salmonella* spp. treated with only hot water showed lower rates of germination, of about 65%. This could denote higher sensitivity of the seeds to this method. Also, the length of the sprouts was not affected overall (data not shown). Hence, these treatments could be applied on the seeds with relatively little impact on germination and growth processes.

The effect of the disinfection treatments on the bacterial load was also evaluated. In this study, the reduction of microorganisms was expected, not only on the seeds but also after germination. Overall, the treatment more suitable for microorganism reduction was disinfection of seeds with SDS followed by Amukine treatment after germination.

In the case of seed decontamination, the protocol that showed better results—less microbiological growth—in *E. coli* contamination was the one used in [4]. Treatment with hot water reduced the growth of *E. coli*, showing that heat inactivation is effective for this pathogen, as also observed elsewhere [36]. On the other hand, *Salmonella* was more susceptible to SDS treatments, which reduced colony count to below the detection levels.

Concerning seed sprouts, the same was verified. For *E. coli* hot water was more effective than SDS as in Barampuram et al. [37] and for *Salmonella* spp. the best results were obtained with a combination of ethanol, sodium hypochlorite and SDS. This could be due to the fact that *Salmonella* spp takes longer to be inactivated by the heat, making treatment with SDS more suitable for this pathogen.

Considering the seeds that suffered treatment before and after germination, Amukine combined with SDS presented a reduction of 1 log for *Salmonella* spp. and 2 logs for *E. coli*, being the better treatment for disinfection. Amukine works by denaturation of the cell's proteins making them inactive while washing the sprouts with water only dilutes the microorganism existent in the plant, making this treatment not enough for proper disinfection.

## 4. Conclusions

Lentil is an affordable source of dietary plant-based protein and other nutrients and consuming it as a sprout has several associated health advantages, as this process induces improved chemical alterations. In the present study, 12 lentil varieties were analysed for their germination capacity and nutritional composition. Germination rates ranged between 72% and 93% and Du Puy, Rosana, Kleine Rote and Kleine Späths II were the highest performing varieties. In terms of protein, Du Puy, Kleine Schwarze, Rosana, Flora, Große Rote and Kleine Späths II had the highest percentages. Of these, Du Puy, Große Rote and Kleine Späths II showed a significant increase in protein concentration after germination. Furthermore, the germination process also impacted micronutrient levels, especially Zn and Mn, that were commonly increased in Du Puy variety. In regards to macronutrients, Ca was increased in Dunkelgrün Marmorierte, Du Puy, Kleine Schwarze and Rosana.

To assess microbiological safety of the sprouts, different disinfection treatments were tested, both for seeds and seed sprouts of Rosana variety. We also concluded that the most effective method for disinfection was the application of 70% ethanol, 1.2% sodium hypochlorite and 0.02% SDS before germination and Amukine afterwards. Seed decontamination for sprout consumption remains a challenge to the sprout industry. Additionally, under common household conditions, seed disinfection is usually neglected, increasing the risk of illness associated with foodborne pathogens in contaminated sprouts.

**Author Contributions:** Conceptualization, C.S.S. and M.W.V.; Methodology, C.S.S.; Validation, M.W.V.; Formal analysis, C.S.S. and B.S.; Investigation, C.S.S. and B.S.; Resources, L.M.P.V., S.G. and M.W.V.; Data curation, C.S.S.; Writing—original draft preparation, C.S.S.; Writing—review and editing, B.S., L.M.P.V., S.G. and M.W.V.; Supervision, C.S.S. and M.W.V.; Project administration, M.W.V.; Funding acquisition, M.W.V. All authors have read and agreed to the published version of the manuscript.

**Funding:** This research received funding from the European Union's Horizon 2020 research and innovation program under grant agreement no. 727973, TRUE project.

**Acknowledgments:** The authors acknowledge funding from the European Union's Horizon 2020 research and innovation program under grant agreement no. 727973, TRUE project; and the scientific support of National Funds from FCT - Fundação para a Ciência e a Tecnologia through project UID/Multi/50016/2019. The authors would also like to thank Michael A. Grusak, Centre Director of the USDA-ARS, for his careful English revision.

**Conflicts of Interest:** The authors declare no conflict of interest.

## References

1. Magkos, F.; Tetens, I.; Bugel, S.G.; Felby, C.; Schacht, S.R.; Hill, J.O.; Ravussin, E.; Astrup, A. A perspective on the transition to plant-based diets: A diet change may attenuate climate change, but can it also attenuate obesity and chronic disease risk? *Adv. Nutr.* **2019**, *11*, 1–9. [CrossRef] [PubMed]
2. Benincasa, P.; Falcinelli, B.; Lutts, S.; Stagnari, F.; Galieni, A. Sprouted grains: A comprehensive review. *Nutrients* **2019**, *11*, 421. [CrossRef] [PubMed]
3. Hoover, R.; Zhou, Y. In vitro and in vivo hydrolysis of legume starches by α-amylase and resistant starch formation in legumes—A review. *Carbohydr. Polym.* **2003**, *54*, 401–417. [CrossRef]

4. Bari, M.L.; Nei, D.; Enomoto, K.; Todoriki, S.; Kawamoto, S. Combination Treatments for Killing *Escherichia coli* O157:H7 on Alfalfa, Radish, Broccoli, and Mung Bean Seeds. *J. Food Prot.* **2009**, *72*, 631–636. [CrossRef] [PubMed]
5. Thavarajah, D.; Thavarajah, P.; Wejesuriya, A.; Rutzke, M.; Glahn, R.P.; Combs, G.F.; Vandenberg, A. The potential of lentil (*Lens culinaris* L.) as a whole food for increased selenium, iron, and zinc intake: Preliminary results from a 3 year study. *Euphytica* **2011**, *180*, 123–128. [CrossRef]
6. Xu, B.; Chang, S.K. Phenolic substance characterization and chemical and cell-based antioxidant activities of 11 lentils grown in the Northern United States. *J. Agric. Food Chem.* **2010**, *58*, 1509–1517. [CrossRef] [PubMed]
7. Khazaei, H.; Subedi, M.; Nickerson, M.; Martínez-Villaluega, C.; Frias, J.; Vandenberg, A. Seed protein of lentils: Current status, progress, and food applications. *Foods* **2019**, *8*, 391. [CrossRef]
8. Siva, N.; Johnson, C.R.; Richard, V.; Jesch, E.D.; Whiteside, W.; Abood, A.A.; Thavarajah, P.; Duckett, S.; Thavarajah, D. Lentil (*Lens culinaris* Medikus) diet affects the gut microbiome and obesity markers in rat. *J. Agric. Food Chem.* **2018**, *66*, 8805–8813. [CrossRef]
9. García-Mora, P.; Martín-Martínez, M.; Bonache, M.A.; González-Múniz, R.; Peñas, E.; Frias, J.; Martinez-Villaluenga, C. Identification, functional grastrointestinal stability and molecular docking studies of lentil peptides with dual antioxidant and angiotensin I converting enzyme inhibitory activities. *Food Chem.* **2017**, *221*, 464–472. [CrossRef]
10. EFSA. EFSA Assesses the Public Health Risk of Seeds and Sprouted Seeds. Available online: https://www.efsa.europa.eu/en/press/news/111115 (accessed on 10 February 2020).
11. Kintz, E.; Byrne, L.; Jenkins, C.; McCarthy, N.; Vivancos, R.; Hunter, P. Outbreaks of Shiga toxin-producing *Escherichia coli* linked to sprouted seeds, salad, and leafy greens: A systematic review. *J. Food Prot.* **2019**, *82*, 1950–1958. [CrossRef]
12. U.S. Food and Drug Administration. *Reducing Microbial Food Safety Hazards in the Production of Seed for Sprouting: Guidance for Industry*; Office of Food Safet: College Park, MD, USA, 2019.
13. Scouten, A.; Beuchat, L. Combined effects of chemical, heat and ultrasound treatments to kill *Salmonella* and *Escherichia coli* O157:H7 on alfalfa seeds. *J. Appl. Microbiol.* **2001**, *92*, 668–674. [CrossRef] [PubMed]
14. CIDRAP. Sprouts Blamed in Big Ontario Salmonella Outbreak. Regents of the University of Minesotta. Available online: http://www.cidrap.umn.edu/news-perspective/2005/12/sprouts-blamed-big-ontario-salmonella-outbreak (accessed on 10 February 2020).
15. Bergspica, I.; Ozola, A.; Miltina, E.; Alksne, L.; Meistere, I.; Cibrovska, A.; Grantina-Ievina, L. Occurence of pathogenic and potentially pathogenic bacteria in microgreens, sprouts, and sprouted seeds on retail market in Riga, Latvia. *Foodborne Pathog. Dis.* **2020**, *17*, 6.
16. Ding, H.; Fu, T.; Smith, M.A. Microbial contamination in sprouts: How effective is seed disinfection treatment? *J. Food Sci.* **2013**, *78*, 495–501. [CrossRef] [PubMed]
17. Standard Operating Procedure (SOP) for Laboratory Disinfection New Orleans LSU Health. 2013. Available online: https://www.lsuhsc.edu/admin/pfm/ehs/docs/decon.pdf (accessed on 4 April 2019).
18. Kottek, M.; Grieser, J.; Beck, C.; Rudolf, B.; Rubel, F. World map of the Köppen-Geiger climate classification updated. *Meteorol. Z.* **2006**, *15*, 259–263. [CrossRef]
19. Shanmugam, V.; Wang, Y.W.; Tsednee, M.; Karunakaran, K.; Yeh, K.C. Glutathione plays an essential role in nitric oxide-mediated iron-deficiency signaling and iron-deficiency tolerance in Arabidopsis. *Plant J.* **2015**, *84*, 464–477. [CrossRef] [PubMed]
20. Santos, C.S.; Roriz, M.; Carvalho, S.M.P.; Vasconcelos, M.W. Iron partitioning at an early growth stage impacts iron deficiency responses in soybean plants (*Glycine max* L.). *Front. Plant Sci.* **2015**, *6*, 325. [CrossRef]
21. GraphPad. Available online: https://www.graphpad.com/scientific-software/prism/ (accessed on 21 March 2020).
22. Xu, M.; Jin, Z.; Simsek, S.; Hall, C.; Rao, J.; Chen, B. Effect of germination on the chemical composition, thermal, pasting, and moisture sorption properties of flours from chickpea, lentil, and yellow pea. *Food Chem.* **2019**, *295*, 579–587. [CrossRef]
23. Santos, C.S.; Carbas, B.; Castanho, A.; Bronze, M.R.; Serrano, C.; Vasconcelos, M.W.; Vaz Patto, M.C.; Brites, C. Relationship between seed traits and pasting and cooking behavior in a pulse germplasm collection. *Crop Pasture Sci.* **2018**, *69*, 892–903. [CrossRef]

24. Santos, C.S.; Carbas, B.; Castanho, A.; Vasconcelos, M.W.; Vaz Patto, M.C.; Bomoney, C.; Brites, C. Variation in pea (*Pisum sativum* L.) seed quality traits defined by physicochemical functional properties. *Foods* **2019**, *8*, 570. [CrossRef]
25. Fouad, A.A.; Rehab, F.M.A. Effect of germination time on proximate analysis, bioactive compounds and antioxidant activity of lentil (*Lens culinaris* Medik.) sprouts. *Acta Sci. Pol. Technol. Aliment.* **2015**, *14*, 233–246. [CrossRef]
26. Márton, M.; Mándoki, Z.; Csapó-Kiss, Z.; Csapó, J. The role of sprouts in human nutrition. A review. *Acta Univ. Sapientiae, Aliment.* **2010**, *3*, 81–117.
27. Zielinska-Dawidziak, M.; Staniek, H.; Król, E.; Piasecka-Kwiatkowska, D.; Twardowski, T. Legume seeds and cereals grains' capacity to accumulate iron while sprouting in order to obtain food fortificant. *Acta Sci. Pol. Technol. Aliment.* **2016**, *15*, 333–338. [CrossRef] [PubMed]
28. Luo, Y.; Xie, W.; Jin, X.; Wang, Q.; He, Y. Effects of germination on iron, zinc, calcium, manganese, and rcopper availability from cereals and legumes. *CyTA J. Food* **2014**, *12*, 22–26. [CrossRef]
29. Plaza, L.; Ancos, B.; Cano, M.P. Nutritional and health-related compounds in sprouts and seeds of soybean (*Glycine max*), wheat (*Triticum aestivum* L.) and alfalfa (*Medicago sativa*) treated by a new drying method. *Eur. Food Res. Technol.* **2003**, *216*, 138–144. [CrossRef]
30. Mwikya, S.M.; Camp, J.V.; Rodriguez, R.; Huyghebaert, A. Effects of sprouting on nutrient and antinutrient composition of kidney beans (*Phaseolus vulgaris* var. *Rose coco*). *Eur. Food Res. Technol.* **2001**, *212*, 188–191.
31. Devi, C.B.; Kushwaha, A.; Kumar, A. Sprouting characteristics and associated changes in nutritional composition of cowpea (*Vigna unguiculata*). *J. Food Sci. Technol.* **2015**, *52*, 6821–6827. [CrossRef]
32. Zerche, S.; Ewald, A. Seed potassium concentration decline during maturation is inversely related to subsequent germination of primrose. *J. Plant Nutr.* **2005**, *28*, 573–603. [CrossRef]
33. Hasanuzzaman, M.; Bhuyan, M.H.M.B.; Nahar, K.; Hossain, M.S.; Mahmud, J.A.; Hossen, M.S.; Masud, A.A.C.; Moumita; Fujita, M. Potassium: A vital regulator of plant responses and tolerance to abiotic stresses. *Agronomy* **2018**, *8*, 31. [CrossRef]
34. U.S. Food and Drug Administration. *Guidance for Industry: Reducing Microbial Food Safety Hazards for Sprouted Seeds*; Office of Food Safety: College Park, MD, USA, 1999.
35. Neurath, H.; Greenstein, J.P.; Putnam, F.W.; Erickson, J.A. The Chemistry of Protein Denaturation. *Chem. Rev.* **1944**, *34*, 157–265. [CrossRef]
36. Theitler, D.; Nasser, A.; Gerchman, Y.; Kribus, A.; Mamane, H. Synergistic effect of heat and solar UV on DNA damage and water disinfection of *E. coli* and bacteriophage MS2. *J. Water Health* **2012**, *10*, 605–618. [CrossRef]
37. Barampuram, S.; Allen, G.; Krasnyanski, S. Effect of various sterilization procedures on the in vitro germination of cotton seeds. *Plant Cell Tissue Org.* **2014**, *118*, 179–185. [CrossRef]

*Article*

# Bioavailability of Melatonin from Lentil Sprouts and Its Role in the Plasmatic Antioxidant Status in Rats

**Miguel Rebollo-Hernanz [1,2], Yolanda Aguilera [1,2], Teresa Herrera [2], L. Tábata Cayuelas [2], Montserrat Dueñas [3], Pilar Rodríguez-Rodríguez [4], David Ramiro-Cortijo [4], Silvia M. Arribas [4] and María A. Martín-Cabrejas [1,2,*]**

1 Department of Agricultural Chemistry and Food Science, School of Science, Universidad Autónoma de Madrid, 28049 Madrid, Spain; miguel.rebollo@uam.es (M.R.-H.); yolanda.aguilera@uam.es (Y.A.)
2 Institute of Food Science Research, CIAL (UAM-CSIC), 28049 Madrid, Spain; teresa.herrera@uam.es (T.H.); cial.foodchem@gmail.com (L.T.C.)
3 Grupo de Investigación en Polifenoles, Unidad de Nutrición y Bromatología, Facultad de Farmacia, Universidad de Salamanca, Campus Miguel de Unamuno, 37007 Salamanca, Spain; mduenas@usal.es
4 Department of Physiology, School of Medicine, Universidad Autónoma de Madrid, 28029 Madrid, Spain; pilar.rodriguezr@uam.es (P.R.-R.); dramiro@bidmc.harvard.edu (D.R.-C.); silvia.arribas@uam.es (S.M.A.)
* Correspondence: maria.martin@uam.es; Tel.: +34 91 001 7900 (ext. 913)

Received: 14 January 2020; Accepted: 10 March 2020; Published: 12 March 2020

**Abstract:** Melatonin is a multifunctional antioxidant neurohormone found in plant foods such as lentil sprouts. We aim to evaluate the effect of lentil sprout intake on the plasmatic levels of melatonin and metabolically related compounds (plasmatic serotonin and urinary 6-sulfatoxymelatonin), total phenolic compounds, and plasmatic antioxidant status, and compare it with synthetic melatonin. The germination of lentils increases the content of melatonin. However, the phenolic content diminished due to the loss of phenolic acids and flavan-3-ols. The flavonol content remained unaltered, being the main phenolic family in lentil sprouts, primarily composed of kaempferol glycosides. Sprague Dawley rats were used to investigate the pharmacokinetic profile of melatonin after oral administration of a lentil sprout extract and to evaluate plasma and urine melatonin and related biomarkers and antioxidant capacity. Melatonin showed maximum concentration (45.4 pg/mL) 90 min after lentil sprout administration. The plasmatic melatonin levels increased after lentil sprout intake (70%, $p < 0.05$) with respect to the control, 1.2-fold more than after synthetic melatonin ingestion. These increments correlated with urinary 6-sulfatoxymelatonin content ($p < 0.05$), a key biomarker of plasmatic melatonin. Nonetheless, the phenolic compound content did not exhibit any significant variation. Plasmatic antioxidant status increased in the antioxidant capacity upon both lentil sprout and synthetic melatonin administration. For the first time, we investigated the bioavailability of melatonin from lentil sprouts and its role in plasmatic antioxidant status. We concluded that their intake could increase melatonin plasmatic concentration and attenuate plasmatic oxidative stress.

**Keywords:** melatonin; bioavailability; lentil sprouts; phenolic compounds; antioxidant status; pharmacokinetics

---

## 1. Introduction

Melatonin (*N*-acetyl-5-methoxytryptamine) is a multifunctional indoleamine acting as a neurohormone and bioactive antioxidant synthesized from L-tryptophan via serotonin in both plants and animals [1]. Melatonin is considered an excellent naturally occurring antioxidant, displaying free radical and reactive species scavenging properties. Melatonin has been shown to detoxify different (i) reactive oxygen species (ROS), e.g., singlet oxygen, hydrogen peroxide, and hydroxyl radicals; reactive nitrogen species (RNS), e.g., peroxynitrite, and peroxynitrous acid, and nitric oxide; and (iii) reactive chlorine species (RCS), such as hypochlorous acid [2]. Additionally, melatonin upregulates the expression and activity of several antioxidant enzymes, preserving the optimal function of mitochondria, and contributing to the attenuation of oxidative stress [3]. Melatonin may be associated with the prevention of chronic diseases related to aging and oxidative stress, including type-2 diabetes, obesity, and cardiovascular diseases [4–7].

Besides the significant influence of the light-dark cycle, nutritional factors such as macro- and micro- nutrient intake, meal timing, and type of diet can also modulate the concentration of melatonin in plasma [8]. The inclusion of melatonin-rich plant foods in the diet may promote human health owing to its biological activities and bioavailability [9,10]. Melatonin is broadly used as a food supplement worldwide, including in the United States and Europe, to counteract physiologically diminished activity of the pineal gland [11]. The European Food Safety Authority (EFSA) has established guidance on regulating melatonin as a dietary supplement, unlike the U.S. Food and Drug Administration (FDA). The EFSA only supports melatonin administration for the treatment of sleeping disorders and jet lag. Moreover, certain European countries have restricted melatonin use to decreasing quantities [12]. Hence, the search for novel natural food sources of melatonin is the object of increasing interest. Melatonin-rich food consumption could influence health by increasing the plasmatic melatonin levels and improving antioxidant status [13].

Melatonin has been found in noteworthy concentrations in teas and herbal infusions [14], and other foodstuffs such as grapes [15], olive oil, cereals, fruits, nuts [16], and legumes [17]. Fermentation food processes increase melatonin concentration, as shown in beer and wine, among others [18]. Germination has been commonly used for the enhancement of legume nutritional value. This processing significantly diminishes the concentration of nonnutritional compounds, enhances the digestibility of proteins, and modifies dietary fiber fractions. Furthermore, germination magnifies the content of bioactive antioxidant compounds, including melatonin [19–21].

The bioavailability of plant-food-derived melatonin has been confirmed, with reported increments in plasmatic melatonin levels upon consumption of enriched melatonin foods [22]. Synthetic melatonin is absorbed in the gastrointestinal tract 45 min post-oral administration, following metabolism in the liver. From 3 to 76% of melatonin consumed reaches the circulation, and it has a short half-life, $t_{1/2}$ (i.e., about 20–40 min) [23]. Then, melatonin is mainly excreted as sulfated metabolites in the urine, primarily as 6-sulfatoxymelatonin (aMT6s) [24]. Research on the bioavailability and pharmacokinetics of dietary melatonin is quite scarce to date. Despite the influence of the food matrix and digestion processes on the absorption of this amphipathic molecule, studies have revealed that the intake of melatonin-rich foods gives rise to an increase in circulating melatonin concentrations [25,26].

Lentils are a source of bioactive compounds, including dietary fiber, protease inhibitors, lectins, saponins, phytosterols, and phenolic compounds. Germination generally increases the content of soluble components. Minor compounds found in lentils and lentil sprouts are considered to be responsible for their health-promoting effects [27]. Lentils have been shown to possess a wide range of biological activities demonstrated not only *in vitro*, but also *in vivo* and by clinical trials; antioxidant, antidiabetic, antiobesity, cardioprotective, and cancer-preventive are the most remarkable properties [28].

Therefore, we aimed to evaluate the influence of the intake of lentil sprouts on the plasmatic concentration of melatonin and metabolically related compounds (plasmatic serotonin and urinary 6-sulfatoxymelatonin), as well as total phenolic compounds and the effects on plasmatic antioxidant status in comparison to synthetic melatonin administration.

## 2. Materials and Methods

### 2.1. Materials

Synthetic melatonin (> 99%), Folin-Ciocalteu reagent, gallic acid (> 96%), $Na_2CO_3$, 3′,6′-dihydroxyspiro[isobenzofuran-1(3H),9′-[9H]xanthen]-3-one (fluorescein), 2,2-azobis (2-methyl-propionamidine)-dihydrochloride, 6-hydroxy-2,5,7,8-tetramethylchroman-2-carboxylic acid (Trolox), 2,4,6-tri(2-pyridyl)-s-triazine (TPTZ), and $FeCl_3$ were obtained from Sigma-Aldrich (St. Louis, MO, USA). All other chemicals and reagents were obtained from VWR (Barcelona, Spain) unless otherwise specified. Melatonin, 6-sulfatoxymelatonin, and serotonin ELISA kits were obtained from IBL-International (Hamburg, Germany). Male Sprague-Dawley rats (3 weeks old) were acquired at the Animal House Facility of the Universidad Autónoma de Madrid (Madrid, Spain). Diet was purchased from Safe (Augy, France). Raw lentils (*Lens culinaris* L., var. Castellana) were provided by the Institute of Food Science, Technology and Nutrition (ICTAN-CSIC, Madrid).

## 2.2. Lentil Sprout Extracts

Lentils were germinated following the procedure described in Aguilera et al. [29]. This process presented good viability, with 96% germination. Lentil sprouts were frozen, freeze-dried, ground, packed in plastic bags, and stored at −20 °C. The analysis of sprouts was carried out in triplicate to determine the melatonin and phenolic compound contents, as well as the *in vitro* antioxidant capacity, as described previously [29]. Melatonin was measured using HPLC-ESI-MS/MS, whereas total phenolic compounds were evaluated using the Folin-Ciocalteu method, and *in vitro* antioxidant capacity was measured by the ORAC method, using spectrophotometric methods; the phenolic profile was characterized suing HPLC-DAD-MS/MS techniques (Figure 1A). The extract from lentil sprouts was prepared by mixing lentil sprout flour (20 g) with ethanol (150 mL) and shaking for 16 h at 4 °C in darkness. The mixture was sonicated for 15 min and filtered with vacuum through 11 µm filters (Whatman). The extract was evaporated at 30 °C to dryness and dissolved in 3 mL phosphate-buffered saline (PBS) buffer to achieve the dose. A synthetic melatonin solution was prepared in PBS to deliver the same dose of the lentil sprouts extract.

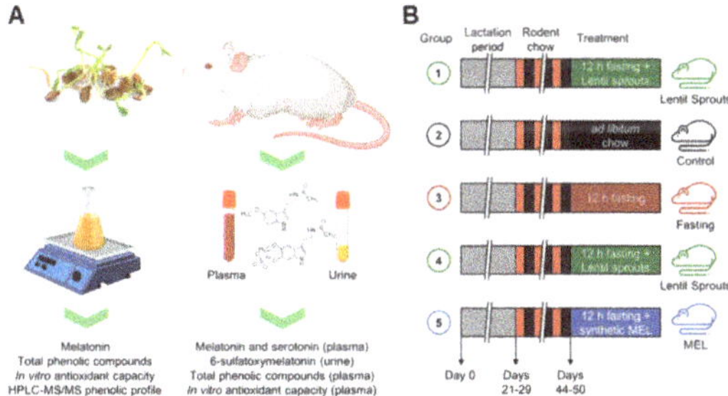


**Figure 1.** Schematic diagram of the experiments design of this research: (**A**) Lentils were germinated for six days, and phytochemicals (melatonin and phenolic compounds) were extracted and characterized, as was the *in vitro* antioxidant capacity of the product. Lentil sprouts were given to rats and plasma and urine were used for the measurement of melatonin, serotonin, 6-sulfatoxymelatonin (immunochemically), and phenolic compounds and the *in vitro* antioxidant capacity following the protocol here illustrated: (**B**) The first group of rats fasted for 12 h. They were then administered a lentil sprout extract via gavage followed sequential blood extraction for the pharmacokinetic study. Groups 2–5 were used in the bioavailability/bioactivity study and subjected to different feeding conditions: ad libitum standard rat chow (Control), 12 h fasting (Fasting), Lentil sprout ingestion after a 12 h fasting (Lentil Sprouts), and the administration of a melatonin solution (MEL) after 12 h fasting, followed in all the cases by blood and urine sampling.

## 2.3. Extraction and Analysis of Phenolic Compounds in Raw Lentils and Lentil Sprouts

### 2.3.1. Extraction of Free and Bound Phenolic Compounds

For free phenolic compound extraction, freeze-dried samples (1 g) were macerated in methanol: HCL (1‰)–water 80:20 (*v/v*) at 4 °C for 16 h. Subsequently, they were centrifuged at 4000 g and 4 °C for 20 min. The extraction process was performed twice. The extracts were combined and concentrated at 30 °C under a vacuum for methanol evaporation. For the extraction of bound phenolics, the residues from the above free phenolic compounds extraction were flushed with $N_2$ and hydrolyzed directly with 10 mL of 4 M NaOH at room temperature for 1 h with shaking. The mixture was acidified to pH 2 with concentrated HCl (8M), centrifuged at 4000 g and 4 °C for 20 min, and extracted three times with 1:1 ethyl ether: ethyl acetate. The organic fractions were evaporated to dryness at 30 °C under a vacuum. For phenolic analysis, the dry extracts were dissolved in 10 mL of water. For purification, an aliquot (4 mL) was passed through a C18 solid-phase extraction (SPE) cartridge (Waters, Milford, MA, USA), and phenolic compounds were eluted with methanol. Afterward, extracts were concentrated under a vacuum in a rotary evaporator (30 °C) and then dissolved in aqueous 0.1% TFA:acetonitrile (90:10 v/v) for phenolic compound analysis. The total content of phenolic compounds was obtained as the sum of free and bound phenolics.

#### 2.3.2. HPLC–DAD–ESI/MS$^n$ qualitative and quantitative analyses of phenolic compounds

Samples were analyzed using a Hewlett-Packard 1100MS (Agilent Technologies, Palo Alto, CA, USA) chromatograph equipped with a quaternary pump, and a diode array detector (DAD) coupled to an HP Chem Station (rev. A.0504) data-processing station. Solvents used were 0.1% formic acid in water (solvent A) and 100% acetonitrile (solvent B). The elution gradient established was 15% B for 5 min, 15–20% B for 5 min, 20–25% B for 10 min, 25–35% B for 10 min, 35–50% B for 10 min, and re-equilibration of the column. The separation of phenolic compounds was performed in a Spherisorb S3 ODS-2 C8 column (Waters, Milford, USA) (3 μm, 150 mm × 4.6 mm i.d.) operating at 35 °C and a flow rate of 0.5 mL/min. Double online detection was carried out in the DAD using 280 nm and 370 nm as preferred wavelengths. A mass spectrometer (MS) connected to the HPLC system via the DAD cell outlet was used and detection was performed in an API 3200 Qtrap (Applied Biosystems, Darmstadt, Germany) equipped with an ESI source, triple quadrupole-ion trap mass analyzer and controlled by the Analyst 5.1 software. The setting parameters were zero grade air as the nebulizer gas (30 psi), turbo gas for solvent drying (400 °C, 40 psi), and nitrogen served as the curtain (20 psi) and collision gas (medium). The quadrupoles were set at unit resolution. The ion spray voltage was set at −4500 V in the negative mode. The MS detector was programmed to perform a series of two consecutive modes: enhanced MS (EMS) and enhanced product ion (EPI) analysis. EMS was employed to show full scan spectra to give an overview of all the ions in the sample. The settings used were declustering potential (DP) −45 V, entrance potential (EP) −6 V, and collision energy (CE) −10 V.

Spectra were recorded in negative ion mode between *m/z* 100 and 1000. EPI mode was further performed to obtain the fragmentation pattern of the parent ion(s) of the previous experiment using the following parameters: DP: −50 V, EP: −6 V, CE: −25 V, and collision energy spread (CES) 0 V. The phenolic compounds were characterized according to their UV and mass spectra and retention times, and comparison with authentic standards when available. For quantitative analyses, calibration curves were prepared by injection of known concentrations of different standard compounds.

### 2.4. *Animals Care and Diet*

Experiments were performed in Sprague Dawley rats. All experimental procedures were approved by the Ethics Review Board of Universidad Autónoma de Madrid and conformed to the Guidelines for the Care and Use of Laboratory Animals (NIH publication No. 85–23, revised in 1996), the Spanish legislation (RD 1201/2005) and the Directive 2010/63/EU on the protection of animals used for scientific purposes. The rats were housed under controlled conditions of 22 °C, 40% relative humidity, and 12/12 h light/dark photoperiod. After weaning (day 21–29), rats were kept for 20–25 days under a reversed light/dark cycle (Figure 1; Table 1). They were fed ad libitum with a breeding diet (SAFE A03) containing 51.7% carbohydrates, 21.4% protein, 5.1% lipids, 3.9% fiber, 5.7% minerals, and 12.2% humidity. Drinking water was sterilized by UV, mechanical, and chemical treatments, and provided ad libitum in all cases. All the animals were housed in buckets on aspen wood bedding, which was replaced once a week. The health and welfare of the animals were monitored by staff at least once a day. Monitoring of animals' health ensured that they were free of any pathogens that may interact with any of the parameters studied.

**Table 1.** Experimental conditions for the pharmacokinetic and bioavailability/bioactivity studies considering the intervention, rat population, date of entrance and time into inverse photoperiod, sampling day, mean group weight on the day of the experiment, and sampling time.

| | Pharmacokinetic Study | Bioavailability and Bioactivity Study | | | |
|---|---|---|---|---|---|
| | | Control | Fasting | Lentil Sprout | MEL |
| Group | 1 | 2 | 3 | 4 | 5 |
| Intervention | 12 h-fast Lentil sprout | None | 12 h-fast | 12 h-fast Lentil sprout | 12 h-fast Melatonin |
| Rat population (N) | 8 ♂ | 10 ♂ | 10 ♂ | 10 ♂ | 10 ♂ |
| Entrance into inverse photocycle (day) | 23 | 21 | 24 | 29 | 29 |
| Time in inverse photocycle (days) | 22 | 25 | 20 | 20 | 21 |
| Sampling day | 45 | 46 | 44 | 49 | 50 |
| Weight (g) | 195 ± 11 | 203 ± 11 | 197 ± 9 | 198 ± 10 | 202 ± 8 |
| Sampling time | 10:00 to 14:00 | 11:00 | 11:00 | 11:00 | 11:00 |

### 2.5. *Pharmacokinetic Study*

In a pharmacokinetic study, rats ($n$ = 8) were orally administered with a single dose (50 µg melatonin/kg) of lentil sprouts extract via gastric gavage. Animals were previously fasted for 12 h but had free access to water before the experiment. Blood samples were withdrawn at 0, 60, 90, 120, 180, and 240 min after dosing through sublingual bleeding and transferred to vials containing 5% heparin, centrifuged at 4 °C for 15 min at 3000 $g$. Pharmacokinetic parameters were calculated using the PKSolver [30]. The area under the curve (AUC) is the integral of the plasma concentration of an altered drug against an interval of definite time. Pharmacokinetic parameters including plasma elimination rate constant ($K_e$), elimination half-life ($t_{1/2}$), time to reach measured maximum plasma concentration ($T_{max}$), measured maximum plasma concentration ($C_{max}$), area under the plasma concentration-time curve from the time zero to the time of last quantifiable concentration ($AUC_{0-t}$), area under the plasma concentration-time curve from time zero to infinity ($AUC_{0-\infty}$), area under the plasma (first) moment concentration-time curve from time zero to infinity ($AUMC_{0-\infty}$), mean residence time (MRT), volume of distribution ($V_D/F$), and clearance ($Cl/F$) were determined by noncompartmental methods.

### 2.6. *Bioavailability and Bioactivity Study*

To measure the bioavailability and biological activity of melatonin from lentil sprouts, rats were grown as previously described until day 44–50. At the end of this period, the rats were subjected to four different feeding conditions ($n$ = 10 per group): standard rat chow without extract (Control), a 12 h-fasting period without extract (Fasting), a 12 h fasting period followed by administration of lentil sprout extract (50 µg melatonin/kg) (Lentil Sprouts), or a 12 h fasting period followed by the administration of 50 µg synthetic melatonin/kg solution (MEL). Each of the four groups (10 rats) was administered the treatment and sacrificed on a different day. Weights were similar on the day of treatments. Having a small number of rats, we were able to diminish the influence of rat photoperiod (changes in melatonin content due to light cycles). The administration was performed through a gavage using a suitable intubation cannula. After that, the rats were individually caged, and a cling film was placed below the cage to obtain the urine samples. After 90 min of extract administration, all rat groups were anesthetized by $CO_2$. Urine was collected from the cling film with a pipette and transferred to a vial. All experiments were carried out at 11 a.m., which was the peak time of melatonin production under the light/dark cycle. The blood was collected by cardiac puncture, transferred to vials containing 5% heparin, and centrifuged at 4 °C for 15 min at 2100 g. The plasma was then divided into aliquots into 1 mL vials and kept frozen at −80 °C to assess several biomarkers related to melatonin metabolism. Before the analysis of melatonin, serotonin, aMT6s, and total phenolic compounds, samples were defrosted on ice in the dark. Phytochemicals and metabolites were isolated using SPE. First, plasma and urine samples were centrifuged at 15000 *g* for 4 min at 4 °C. Then, samples were filtered using C18 SPE cartridges (Waters, Cerdanyola del Vallès, Spain). Finally, the eluted obtained fractions were evaporated to dryness (Speed Vac SC 200, Savant Instruments, Farmingdale, NY, USA) and reconstituted before being assayed. According to the volume obtained, the analyses were carefully designed to maximize the use of biological samples; aliquots were employed to measure melatonin, 6-sulfatoxymelatonin, serotonin, total phenolics, and the *in vitro* antioxidant capacity.

### 2.7. Melatonin Determination

Plasma residues were dissolved in Milli-Q water, and melatonin levels were determined by a competitive enzyme immunoassay kit (Catalog No. RE54021, IBL) according to the manufacturer's guidelines. The kit is characterized by an analytical sensitivity of 1.6 pg/mL and high analytical specificity (low cross-reactivity).

### 2.8. 6-Sulfatoxymelatonin (aMT6s) Determination

Urine residues were dissolved in tris-buffered saline (TBS) and protected from direct sunlight. aMT6s levels were determined by a competitive enzyme immunoassay kit (Catalog No. RE54031, IBL) according to the manufacturer´s instructions. The assay sensitivity was 1.0 ng/mL.

### 2.9. Serotonin Quantification

Plasma residues were dissolved in Milli-Q water, and serotonin levels were determined by a competitive enzyme immunoassay kit (Catalog No. RE59121, IBL) according to the manufacturer´s guidelines. The kit is characterized by an analytical sensitivity of 2.68 ng/mL and high analytical specificity (low cross-reactivity).

### 2.10. Total Phenolic Compounds (TPC)

Plasma residues were dissolved in Milli-Q water, and total phenolic compounds were determined by Folin-Ciocalteu colorimetric method, according to Singleton et al. [31] using gallic acid as standard. Samples (10 μL) were added to 150 μL of Folin-Ciocalteu reagent (diluted 1:15, *v/v* in Milli-Q water). After exactly 3 min, 20% $Na_2CO_3$ (50 μL) was added to each well. After 120 min at room temperature, the absorbance of the blue complex was measured at 760 nm using a microplate reader. The concentration of total phenolic compounds was expressed as mg GAE/mL.

### 2.11. ORAC (Oxygen Radical Absorbance Capacity)

The above plasma samples were used to determine the radical scavenging activity via the ORAC method using fluorescein as a fluorescence probe [14]. The assay was performed at 37 °C in 75 mM phosphate buffer (pH 7.4). The reaction mixture (200 μL) contained fluorescein (70 nM), 2,2-azobis (2-methyl-propionamidine)-dihydrochloride (12 mM), and samples (Trolox or samples). Fluorescence was read at 485/520 nm, excitation/emission, respectively. Black 96-well untreated microplates (PS Balck, Porvair, Leatherhead, UK) were used. The fluorescence signal was measured every minute for 80 min after automatic shaking. All reaction mixtures were prepared in triplicate, running them at least three times independently. Fluorescence was normalized to the oxidation control (phosphate buffer) and stability control (no antioxidant). The ORAC values were expressed as mM Trolox equivalents (mM TE) according to the protocol previously described [14].

### 2.12. FRAP (Ferric Reducing Antioxidant Power)

The plasma samples were also used to assess the reducing antioxidant potential using the FRAP assay, as previously described [14]. Briefly, 300 μL of a working FRAP reagent [acetate buffer 0.3 M pH 3.6, 10 mM TPTZ in 40 mM HCl and 20 mM $FeCl_3 \cdot 6H_2O$ (10:1:1) (*v/v/v*)] were tempered to 37 °C, and then 10 μL of plasma samples were added. The absorbance was taken at 593 nm against reagent blank after 10 min. FRAP values were calculated and reported as μM Trolox equivalents (μM TE).

### 2.13. Statistical Analysis

Each sample was analyzed in triplicate. Data were reported as mean ± standard deviation (SD). The data were analyzed using the *T*-test or by one-way analysis of variance (ANOVA) and *post hoc* Tukey test. Relationships between the analyzed parameters were evaluated by computing Pearson linear correlation coefficients setting the level of significance at $p < 0.05$, $p < 0.01$, and $p < 0.001$. The statistical analysis was performed by SPSS 23.0.

## 3. Results and Discussion

### 3.1. Lentil Sprouts are a Source of Melatonin and Antioxidant Phytochemicals

Lentil sprouts germinated for 6 days exhibited high amounts of melatonin (1.01 μg/g) and phenolic compounds (TPC accounted for 3.47 mg/g) (Table 2). The melatonin concentration underwent a drastic increase (more than 2000-fold) compared to that of raw lentils. Germination induced a significant ($p < 0.05$) increment in the content of this indoleamine, which has also been seen in other legumes and sprouted vegetables [17]. The concentration of melatonin in these germinated lentils is considered higher than the content in other plant foods [32]. Melatonin has also been found in outstanding levels in fruits, coffee, and its by-products, or fermented foods like wine and beer [33,34]. Nonetheless, the concentration of phenolic compounds (TPC) exhibited a slight decrease (28%), mainly due to the loss of FPC (30% reduction). The proportion free phenolics:bound phenolics (FPC: BPC) remained unchanged. This reduction has also been reported by other authors, not only in lentils [35], but also in beans or chickpeas [36]. The decrease of free phenolics during germination can be associated with their capacity for scavenging ROS in the plant cell during germination. Likewise, phenoloxidase and peroxidase enzymes, triggered during germination, might cause a diminution in the concentration of free phenolics [37].

**Table 2.** Melatonin content, free (FPC), bound (BPC), and total phenolic compounds (TPC), FPC: BPC ratio, and *in vitro* antioxidant capacity measured by the ORAC method in raw lentil and lentil sprouts and the equivalence of the dose of lentil sprouts given to rats.

| | Melatonin (μg) | FPC (mg GAE) | BPC (mg GAE) | TPC (mg GAE) | FPC: BPC (ratio) | ORAC (μmol TE) |
|---|---|---|---|---|---|---|
| Raw lentils (per g) | (0.46 ± 0.06) · $10^{-3}$ | 4.50 ± 0.17 | 0.38 ± 0.02 | 4.88 ± 0.19 | (12 ± 1):1 | 20.24 ± 1.54 |
| Lentil sprouts (per g) | 1.01 ± 0.09 *** | 3.15 ± 0.21 ** | 0.32 ± 0.03 * | 3.47 ± 0.24 ** | (10 ± 2):1 | 77.23 ± 5.78 *** |
| Dose (per kg) | 50.0 | 155.9 | 51.0 | 171.8 | - | 52.0 |

Results are reported as mean ± SD ($n = 3$). Mean values followed by superscript asterisks significantly differ (raw vs. sprouted lentils) when subjected to *T*-test (* $p < 0.05$, ** $p < 0.01$, *** $p < 0.001$)

Because of the changes produced by germination, the antioxidant capacity of the germinated lentils was higher (3.8-fold) in contrast to the potential of raw lentils. The *in vitro* antioxidant capacity of this product can be considered high in comparison with other legumes or plant-foods [38]. Germination has been shown to be a cost-effective and sustainable strategy to increase the antioxidant capacity of legumes [39]. Therefore, germinated lentils were selected to evaluate melatonin bioavailability and their influence on the plasmatic antioxidant status due to their higher content in melatonin and their higher *in vitro* antioxidant capacity.

Germination Decreased the Concentration of Phenolic Acids and Flavan-3-ols, Preserving Flavonols

A more comprehensive analysis of the antioxidant phytochemicals found in raw lentils and lentil sprouts demonstrated a profound modification in the phenolic profiles of these products (Table 3). The retention time ($R_t$), wavelengths of maximum absorption in the visible region ($\lambda_{max}$), molecular ion, and fragment ion pattern allowed us to identify a total of twenty-two phenolic compounds, including hydroxybenzoic and hydroxycinnamic acids, catechins and procyanidins (flavan-3-ols), and flavonols and flavanones. Major free phenolic compounds in raw lentils were catechins and procyanidins (56%) (Figure 2, Table 3), whereas in the bound fraction, a high proportion of hydroxybenzoic acids (35%) was also present together with catechins (45%). The germination of lentils produced a significant loss of phenolic (hydroxybenzoic and hydroxycinnamic) acids ($p < 0.01$) and flavan-3-ols ($p < 0.001$) in the free phenolic fraction. In the bound phenolic fraction, the decreases were more remarkable for catechins (65% reduction, $p < 0.001$), leading to an increase in the proportion of hydroxybenzoic acids (1.4-fold increase). Concerning the load of total phenolics, the only differences were found for the proportion of flavonols and catechins; catechins decreased (from 55 to 41%) while flavonols increased (from 34 to 48%).

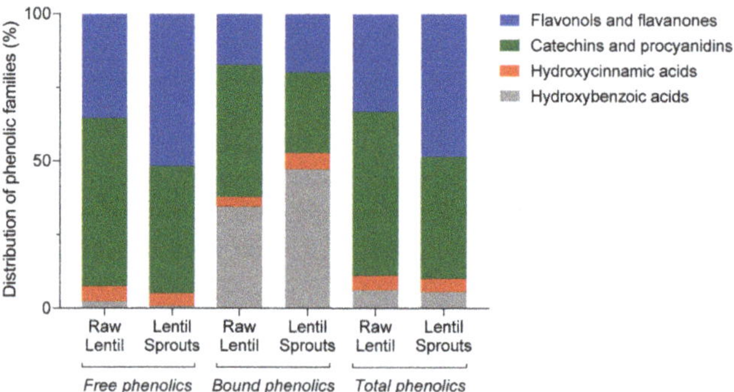


**Figure 2.** Distribution of different phenolic groups (hydroxybenzoic acids, hydroxycinnamic acids, catechins and procyanidins, and flavonols and flavanones) in the free, bound, and total phenolic fractions of raw and germinated lentils.

**Table 3.** Retention time ($R_t$), wavelengths of maximum absorption in the visible region ($\lambda_{max}$), mass spectral data, tentative identification, and quantification (μg/g dry matter) of free and bound phenolic compounds present in raw lentils and lentil sprouts.

| Compounds | $R_t$ (min) | $\lambda_{max}$ (nm) | Molecular ion [M-H]⁻ (*m/z*) | MS² (*m/z*) | Concentration (μg/g) | | | |
|---|---|---|---|---|---|---|---|---|
| | | | | | *Free phenolics* | | *Bound phenolics* | |
| | | | | | Raw lentil | Lentil sprout | Raw lentil | Lentil sprout |
| ***Hydroxybenzoic acids*** | | | | | | | | |
| ***p*-Hydroxybenzoic-hexoside** | **5.48** | 255 | 299 | – | 8.14 ± 4.16 | 3.34 ± 0.20 | 34.32 ± 6.84 | 27.06 ± 1.58 |
| Protocatechuic acid | 6.70 | 257, 294 | 153 | – | 23.22 ± 2.52 | 4.36 ± 0.60$^{***}$ | 26.34 ± 3.00 | 21.48 ± 10.92 |
| **Total** | | | | | **31.36 ± 6.68** | **7.70 ± 0.80$^{**}$** | **60.66 ± 9.84** | **48.54 ± 12.50** |
| ***Hydroxycinnamic acids*** | | | | | | | | |
| *trans-p*-Coumaric derivative | 10.04 | 309 | 417 | – | 18.94 ± 2.08 | 5.16 ± 1.08$^{***}$ | nd | nd |
| *p*-Coumaroyl malic acid | 12.46 | 308 | 278 | 163 | 9.50 ± 1.06 | 11.10 ± 2.88 | nd | nd |
| *p*-Coumaroyl glycolic acid | 12.80 | 314 | 220 | 163 | 12.48 ± 2.32 | 13.68 ± 2.16 | nd | nd |
| *p*-Coumaric acid derivative | 13.10 | 312 | – | 163 | 13.08 ± 0.78 | nd$^{***}$ | nd | nd |
| *trans-p*-Coumaric acid | 18.20 | 314 | 163 | – | 12.72 ± 0.00 | 8.78 ± 1.46$^{**}$ | 2.58 ± 0.00 | 2.64 ± 0.62 |
| *trans*-Ferulic acid | 20.4 | 322 | 193 | – | nd | nd | 3.18 ± 0.34 | 3.16 ± 1.04 |
| **Total** | | | | | **66.72 ± 6.24** | **38.72 ± 7.58$^{**}$** | **5.76 ± 0.34** | **5.80 ± 1.66** |
| ***Catechins and procyanidins*** | | | | | | | | |
| (+)-Catechin | 10.90 | 280 | 289 | 245, 203, 161 | 262.98 ± 37.20 | 150.06 ± 14.08$^{**}$ | 78.74 ± 14.54 | 27.94 ± 5.22$^{***}$ |
| (+)-Catechin 3-*O*-hexoside | 9.02 | 280 | 451 | 289 | 298.58 ± 15.76 | 206.32 ± 5.26$^{***}$ | nd | nd |
| Dimer prodelphinidin | 7.76 | 276 | 593 | 447, 441, 423, 305, 289 | 120.20 ± 2.54 | nd$^{***}$ | nd | nd |
| Dimer procyanidin | 10.40 | 279 | 577 | 289 | 52.68 ± 0.02 | 23.86 ± 2.00$^{***}$ | nd | nd |
| **Total** | | | | | **734.44 ± 55.52** | **380.24 ± 19.34$^{***}$** | **78.74 ± 14.54** | **27.94 ± 5.22$^{***}$** |
| ***Flavonols and flavanone*** | | | | | | | | |
| Kaempferol glucuronide dihexoside | 10.12 | 346 | 785 | 285 | nd | 2.56 ± 0.60$^{**}$ | nd | nd |
| Kaempferol dirutinoside | 14.05 | 346 | 901 | 755, 593, 285 | 408.60 ± 3.72 | 305.74 ± 10.04$^{***}$ | 22.70 ± 4.72 | 8.44 ± 0.46$^{**}$ |
| Kaempferol rutinoside hexoside (I) | 15.00 | 345 | 755 | 593, 285 | 15.02 ± 0.26 | 8.92 ± 1.26$^{**}$ | 1.82 ± 0.12 | 1.90 ± 0.36 |
| Kaempferol rutinoside hexoside (II) | 17.04 | 346 | 755 | 593, 285 | nd | 7.10 ± 0.24$^{***}$ | nd | nd |
| Kaempferol rutinoside rhamnoside (I) | 18.30 | 346 | 739 | 593, 285 | 12.68 ± 0.04 | 16.10 ± 1.00$^{**}$ | 0.82 ± 0.06 | 1.24 ± 0.26 |
| Kaempferol rutinoside rhamnoside (II) | 19.14 | 346 | 739 | 593, 285 | nd | 11.32 ± 3.08$^{**}$ | nd | nd |
| Quercetin 3-glucoside | 20.1 | 356 | 463 | 301 | 3.30 ± 0.26 | 3.02 ± 0.32 | nd | nd |
| Kaempferol glucuronide | 25.40 | 346 | 461 | 285 | 11.22 ± 2.48 | 14.10 ± 1.24 | nd | nd |
| Kaempferol rhamnoside | 34.10 | 346 | 431 | 285 | 10.52 ± 2.06 | 69.00 ± 8.54$^{***}$ | 4.84 ± 0.46 | 8.76 ± 1.36$^{***}$ |
| Eriodyctyol hexoside | 16.30 | 288, 338 (sh) | 449 | 287 | 8.92 ± 1.26 | 18.78 ± 0.96$^{***}$ | nd | nd |
| **Total** | | | | | **470.26 ± 10.08** | **456.64 ± 27.28** | **30.18 ± 5.36** | **20.34 ± 2.44$^{*}$** |
| ***Total phenolic compounds*** | | | | | ***1302.78 ± 78.52*** | ***883.30 ± 55.00$^{**}$*** | ***175.34 ± 30.08*** | ***102.62 ± 21.82$^{*}$*** |

Results are reported as mean ± SD ($n$ = 3). Mean values followed by superscript asterisks significantly differ (raw vs. sprouted lentils) when subjected to *T*-test ($^{*}$ $p < 0.05$, $^{**}$ $p < 0.01$, $^{***}$ $p < 0.001$). nd: nor detected.

The main free phenolic compounds found in raw lentils were kaempferol dirutinoside, (+)-catechin 3-*O*-hexoside, and (+)-catechin (408.6, 298.6, and 263.0 μg/g, respectively) (Table 3). These compounds underwent a substantial reduction during germination (25, 31, and 43%, respectively), leading up to increases in the concentrations of other kaempferol derivatives. During germination, glucosidase activity is enhanced, resulting in a loss of the glycosidic forms of phenolic compounds [40]. β-Glycosidases, upregulated during germination, catalyzes the hydrolysis of β-glycosidic di- and other glycoside conjugates from phenolics, thus releasing both the sugar moiety and the aglycone [40]. Here, we observed an increase in the concentration of both kaempferol rutinoside rhamnoside isomers (2.2-fold) and primarily kaempferol rhamnoside (6.6-fold), produced from the elimination of one glucose or one rutinose (6-O-rhamnosil-glucose) molecules, respectively. Likewise, germination brought about a decrease in the concentration of procyanidin and prodelphinidin dimers, which was also observed by Lopez et al. [41]. Concerning bound phenolics, besides the losses in catechin due to germination, a slight decrease (33%, $p < 0{,}05$) was shown in the content of total flavonols, mainly due to the reduction in the concentration of kaempferol dirutinoside ($p < 0.01$) even though the content of kaempferol rhamnoside increased 1.8-fold ($p < 0.0001$). The content of total phenolics significantly diminished (33%, $p < 0.001$), following a similar trend to that observed for the TPC content. Besides that, the content of total phenolic compounds measured using HPLC-DAD-MS/MS strongly correlated ($r = 0.9980$, $p < 0.001$) with the values obtained using the colorimetric Folin-Ciocalteu method, validating the use of this spectrophotometric method as a screening technique for the analysis of phenolics during different processes [42].

### 3.2. Bioavailability of Melatonin from Lentil Sprouts and its Correlation with Plasma Antioxidant Capacity

A pharmacokinetic study was performed using eight male rats after oral administration of lentil sprouts (50 μg melatonin/kg). Figure 3 shows the mean plasma concentration-time profiles of melatonin, serotonin, and the plasmatic antioxidant capacity measured by the FRAP method. Melatonin (Figure 3A) was detected in basal levels at time 0, following an increase in the concentration (45.4 pg/mL) until 1.5 h postadministration of the lentil sprouts extract. The concentration diminished from that point to achieve basal concentration 4 h after lentil sprouts intake. Regarding plasmatic serotonin concentration (Figure 3B), a significant increase was observed between 60–120 min after the consumption of lentil sprouts. Afterward, a pronounced decay in serotonin concentration was shown (28% at 180 min, 79% at 240 min). The antioxidant status of the plasma (Figure 3C) displayed similar behavior to melatonin. Increases in the antioxidant capacity were detected 60–90 min after the administration of lentil sprouts; then, a slight decrease was perceived before returning to the basal level after 4h. Melatonin concentration in the plasma significantly correlated with the antioxidant capacity ($r = 0.989$, $p < 0.001$). No significant ($p > 0.05$) correlation was detected with serotonin. Melatonin has been shown to reach maximum concentration between 20 to 100 min after its oral administration [43]; similar behavior was observed here. The enhancement of the plasmatic antioxidant capacity (measured by the FRAP method) seemed to be consistent with the antioxidant activity of melatonin. Melatonin was detected after the administration of synthetic melatonin (5 mg/kg) in Sprague-Dawley rats in much higher concentrations (1–2 μg/mL), showing that the co-ingestion with phenolic compounds (caffeic acid or quercetin) could increase melatonin bioavailability [44]. Therefore, the phenolic content (mainly represented by kaempferol glycosides, as previously observed in Table 3) present in the extract might not have a negative impact on melatonin bioavailability. Nevertheless, this hypothesis should be validated in future experiments.

Pharmacokinetic parameters were calculated from the plasmatic melatonin concentration–time profile (Table 4). After oral administration of lentil sprouts (50 μg/kg), melatonin displayed rapid absorption ($T_{max}$ = 90 min) followed by distribution and disposition in rat. The plasmatic half-life was around 100 min, indicating rapid metabolism. Elimination rate constant ($K_e$) values were in the range of results previously obtained [44]. Indeed, melatonin mean residence time (MRT) was low (≈ 200 min). The low doses of melatonin (as such assayed in this study, 50 μg/kg) gave rise to lower

MRT values. Thus, melatonin absorbed from foods (also low doses) only stays in the organism for short periods. The volume of distribution ($V_D/F$) of melatonin was certainly high (0.85 L/g), implying a high distribution of the molecule into the tissues. The systemic plasma clearance ($Cl/F$) of melatonin was 329.9 L/h/kg, which was 100-fold the rat hepatic blood flow rate (3.3 L/h/kg). Melatonin is rapidly metabolized in the liver by cytochrome P450 enzymes, and then sulfated or glucuronidated. Melatonin can be degraded by nonenzymatic pathways involving radical species [45]. These results prove the rapid absorption and metabolism of melatonin from lentil sprouts after acute intake.

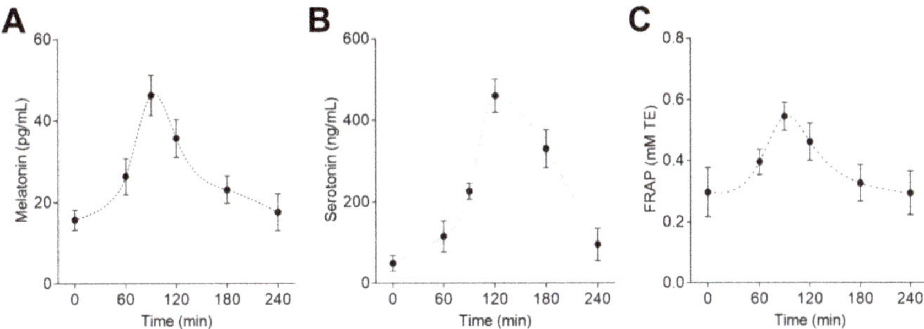


**Figure 3.** Plasma concentration-time profiles of melatonin (**A**), serotonin (**B**), and antioxidant capacity measured by the FRAP method (**C**) after the administration of lentil sprouts (equivalent to 50 µg/kg melatonin). Each point represents the mean ± SD ($n$ = 8).

**Table 4.** Pharmacokinetic parameters of melatonin in rats after oral administration of lentil sprouts.

| Parameter | Oral lentil sprouts (50 µg melatonin/kg) |
|---|---|
| $T_{max}$ (min) | 90.0 ± 0.0 |
| $C_{max}$ (pg/mL) | 45.4 ± 3.5 |
| $K_e$ (1/h) | 0.39 ± 0.05 |
| $t_{1/2}$ (min) | 108.9 ± 13.0 |
| AUC $_{0\text{-}t}$ (pg/mL min) | 6560 ± 963 |
| AUC $_{0\text{-}t}$ /dose (pg/mL min)/(µg/kg) | 131.2 ± 19.3 |
| AUC $_{0\text{-}\infty}$ (pg/mL min) | 9381 ± 2008 |
| AUC $_{0\text{-}t}$ /AUC $_{0\text{-}\infty}$ | 0.71 ± 0.05 |
| AUMC $_{0\text{-}\infty}$ (pg/mL min$^2$) | 1897356 ± 589030 |
| MRT $_{0\text{-}\infty}$ (min) | 199.4 ± 19.8 |
| $V_D/F$ (µg/kg)/(pg/mL) | 0.85 ± 0.08 |
| $Cl/F$ (µg/kg)/(pg/mL)/h | 0.33 ± 0.07 |

$T_{max}$: time to the maximal plasmatic concentration; $C_{max}$: maximal plasmatic concentration; $K_e$: *elimination rate constant;* $t_{1/2}$: elimination half-life; MRT: mean residence time; $V_D/F$: volume of distribution; $Cl/F$: clearance

### *3.3. Lentil Sprouts Intake Greater Elevate Melatonin Biomarkers than the Synthetic Hormone*

Based on the pharmacokinetic results, we compared plasmatic melatonin levels among four different feeding conditions (Control, Fasting, Lentil Sprouts, and MEL) 90 min after treatment (Figure 1). Plasmatic melatonin suffered a decrease (42%) when rats were subjected to a 12 h fasting period (Figure 4A). Periods of fast or caloric restriction have been associated with decreases in plasmatic melatonin concentrations [46]. Nevertheless, the literature shows discrepancies in the role of intermittent fasting and variations in the delivery of plasmatic melatonin [47]. The intake of lentil sprouts significantly ($p < 0.05$) increased melatonin concentration (1.7-fold in contrast to control; 2.9-fold in comparison to the Fasting group). The MEL group also exhibited a meaningful boost in melatonin plasmatic concentration, i.e., 22% lower than that of the Lentil Sprouts group. These outcomes extend the feasibility of consuming lentil sprouts as the melatonin food source. A significant part of the melatonin present in the diet can be absorbed by the gastrointestinal tract, thereby diminishing its levels in plasma. Numerous aspects may be associated with the plasmatic concentration of melatonin, including the age of the animals, the quantity of melatonin consumed, or the administration period [48]. Likewise, the intake of tryptophan has been associated with increased melatonin concentration in plasma [49]. Lentils are considered a source of tryptophan, both protein-bound and free [50]. Lentil sprout intake led to comparable increases of melatonin plasmatic concentration as those reported after walnut ingestion [51]. Fruit consumption can lead to increases in the plasmatic melatonin concentration [52], as can the intake of wine and beer [53]. The bioavailability of phytochemicals is distinctly influenced by the complexity of the food matrix, comprising plant cell-wall polysaccharides, dietary soluble, and insoluble fibers, among other components [54].

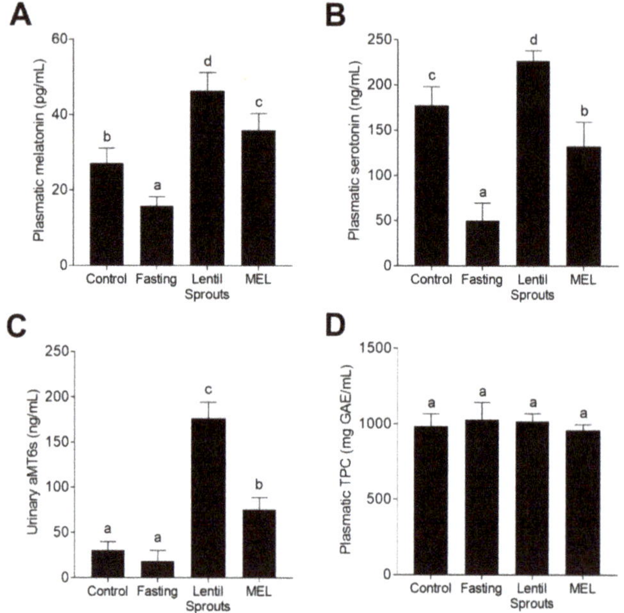


**Figure 4.** The concentration of melatonin (**A**), serotonin (**B**), 6-sulfatoxymelatonin (aMT6s) (**C**), and TPC (**D**) in the plasma and urine samples followed the four experimental conditions (Control, Fasting, Lentil Sprouts, and MEL). The results are expressed as mean ± SD ($n = 10$). Bars with different letters significantly ($p < 0.05$) differ according to ANOVA and Tukey's multiple range test.

Additionally, the concentration of serotonin was studied (Figure 4B), observing a dramatic reduction when rats were fasted, whereas the treatment with the lentil sprout extract enhanced serotonin concentration (4.5-fold in contrast to the Fasting group) as well as the MEL group (2.7-fold). Since serotonin is a hormone related to digestion processes, the 12 h fasting period reduced (72%) the plasmatic concentration. Serotonin is mainly produced by enteroendocrine cells in the gastrointestinal tract in response to a variety of stimulants (nutrients, odorants, and phytochemicals, among others) [55]. Evidence shows that enterochromaffin cells secrete serotonin upon nutrient triggering [56,57]. Serotonin can, however, also be released as an adaptation to long-lasting fasting [58]. As formerly stated, the content of tryptophan from plant sources may also increase the production of extra-pineal melatonin through the metabolization via serotonin to melatonin. Both serotonin and melatonin are primarily synthesized (> 90%) by the gastrointestinal tract, from tryptophan and/or its derivatives, comprising a primary extrapineal pool of these molecules [59]. In fact, serotonin and melatonin can be synthesized not only by animal cells, but also by the microbiota present in the gastrointestinal tract [60].

Urinary aMT6s (Figure 4C), the main melatonin metabolite excreted in urine, did not suffer significant ($p > 0.05$) during the fasting period, while a substantial increase in its concentration was detected upon the administration of both lentil sprouts and melatonin solution (5.8- and 2.5-fold, respectively; $p < 0.05$). The concentration of this melatonin biomarker in urine was associated with the plasmatic concentration of melatonin ($r = 0.926$, $p < 0.05$). After intake, melatonin is promptly transformed in 6-hydroxymelatonin, being further conjugated, giving rise to aMT6s [61]. The concentration of aMT6s excreted in urine after lentil extract and melatonin intake confirmed the highly efficient metabolism of melatonin in rats [62]. The consumption of several fruits and vegetables previously evidenced the bioavailability of melatonin observed as an increase in the concentration of urinary aMT6s [52,63,64]. Nonetheless, the differences observed between Lentil Sprouts and MEL groups should be emphasized. Upon the intake of equal quantities of melatonin, the excretion of aMT6s was significantly ($p < 0.05$) different. Differences in the intake of tryptophan might be the cause of these different results. Tryptophan intake is associated with increased aMT6s production [65]. Tryptophan metabolization in the gastrointestinal tract could increase not only the synthesis of serotonin, but the final production of melatonin and the subsequent urinary excretion of aMT6s [66–68]. Hence, melatonin bioavailability depends not only on the initial concentration of melatonin in food and the matrix effect influencing its absorption, but also on the content of tryptophan which could be converted into melatonin and be part of the plasmatic melatonin load.

The concentration of TPC in the plasma of rats subjected to the four different treatments (Figure 4D) did not show any significant ($p > 0.05$) change in the 90 min posttreatment. The plasmatic $t_{1/2}$ of phenolic compounds frequently ranges 2 to 8 h, but it can reach up to 12–24 h for the larger phenolic structures [69]. In the present work, the levels of phenolic compounds in the plasma of 12 h-fasted rats did not show any fluctuations. The slow release of phenolics from food complex matrices such as regular rat chow, mainly based on vegetable food sources (wheat, corn, wheat bran, barley, soybean, among others) may be the cause of nonsignificant ($p > 0.05$) changes in the fasting group, as well as in the Lentil Sprouts or MEL groups [70]. The occurrence of phenolic compounds along the gut, due to their slow release from foods, appears to be the primary reason for the preservation of the total phenolic load in plasma [71]. Since gut microbiota mediates the synthesis of phenolic acids from larger phenolic compound polymers, such as proanthocyanidins, tannins, and glycosides, via ring fission and oxidation, and glycoside hydrolysis, the resulting phenolic metabolites can be absorbed and reach the systemic circulation [72,73]. Complex phenolic compounds and glycosides need to be metabolized in the colon to be absorbed. The main phenolic compounds composing the melatonin-rich extract administered, primarily kaempferol and (+)-catechin glycosides, may need these metabolization processes to reach their absorbable form and be bioavailable. Thus, the acute intake of lentil sprouts did not result in TPC modification, since phenolic compounds could not reach a digestion stage in which they were bioavailable.

### 3.4. Administration of Lentil Sprouts Enhances Plasmatic Antioxidant Status

Since the major bioactivity of melatonin is its antioxidant potential, we measured the antioxidant status of the plasma by FRAP and ORAC methods 90 min after the treatments to establish the influence of lentil sprout intake (Figure 5). Even if the antioxidant capacity, measured by the FRAP method (Figure 5A), showed no significant ($p > 0.05$) changes during fasting, a 34% increase was observed in the Lentil Sprouts group and a 4% in MEL group. Similarly, the antioxidant capacity measured with the ORAC method showed a similar behavior (Figure 5B). Fasting did not cause significant ($p > 0.05$) alterations in the antioxidant capacity. However, the administration of lentil sprouts gave rise to a 1.8-fold increase; synthetic melatonin produced a less pronounced increment (32% in composition to the fasting group). The antioxidant capacity correlated with the concentration of melatonin (FRAP, $r = 0.977$, $p < 0.05$; ORAC, $r = 0.962$, $p < 0.05$) as well as with the concentration of urinary aMT6s (FRAP, $r = 0.928$, $p < 0.05$; ORAC, $r = 0.991$, $p < 0.01$). The antioxidant capacity did not display any significant correlation ($p > 0.05$) with the concentration of TPC in the plasma. Phenolics might have been slowly released, exhibiting different pharmacokinetics than melatonin. As previously observed, the intake of foodstuffs rich in melatonin such as fruits and derivates (cherries, grapes, and their juices), or fermented beverages (wines and beers) resulted in an enhancement of the plasmatic antioxidant capacity [25,26,53,74]. However, these results may involve a conservative evaluation of antioxidant capacity and should not be attributed solely to the presence of melatonin. These results could be related to other phytochemicals not measured, such as vitamin C or other antioxidants that could be associated with the increase in the plasmatic *in vitro* antioxidant capacity [75]. Furthermore, FRAP and ORAC assays exclusively asses the free radical scavenging capacity. Melatonin displays direct and indirect antioxidant properties, covering not only free radical scavenging capacity but the stimulation of endogenous antioxidant enzyme expression [76]. Melatonin administration has been correlated to the enhancement of the plasmatic antioxidant capacity in both animal models and human studies [77].

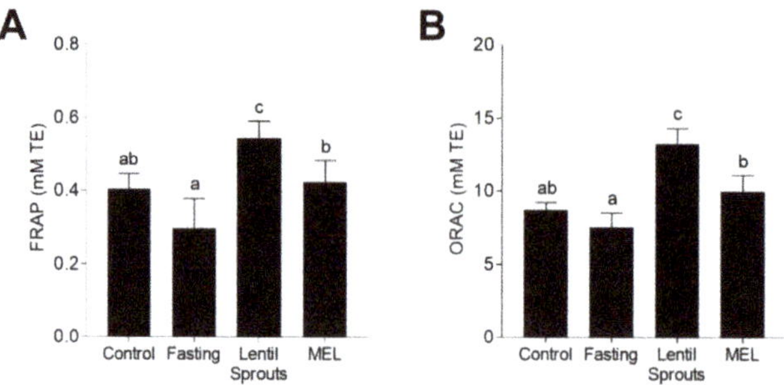


**Figure 5.** Plasmatic antioxidant capacity measured by FRAP (**A**) and ORAC (**B**) methods followed the four experimental conditions (Control, Fasting, Lentil Sprouts, and MEL). The results are expressed as mean ± SD ($n$ = 10). Bars with different letters significantly ($p < 0.05$) differ according to ANOVA and Tukey's multiple range test.

In general, melatonin intake is associated with diminished oxidative stress [78]. The supplementation with melatonin can be beneficial, beyond oxidative stress, in the reduction of inflammation, hypertension, and other metabolic syndrome-related markers [77]. Forthcoming studies will unravel the effects of lentil sprout intake on the gastrointestinal metabolism of melatonin and its influence on the oxidative status, both in plasma and in central tissues where oxidative stress is the primary cause of dysfunction, aging, and the development of chronic diseases. The synthesis of melatonin by the colonic microbiota presents alternative perspectives from which to understand how the content of melatonin increases after the intake of food. The gut microbiota could be considered as a supplementary source of extrapineal melatonin, produced by diverse types of bacteria, deserving future investigations [79].

## 4. Conclusions

In summary, the intake of lentil sprouts increased the concentration of melatonin in the plasma of rats as well as the urinary levels of 6-sulfatoxymelatonin. Likewise, the plasmatic antioxidant capacity was augmented. This study presents the composition of raw lentils and sprouts, and an evaluation of melatonin pharmacokinetics, bioavailability, and biological activity *in vivo* in Sprague Dawley rats. In view of the results, lentil sprouts can be considered as a source of bioavailable melatonin, whose intake could modulate plasmatic oxidative stress and defend the organism from aging and related diseases, among other potential health benefits. Future studies should be conducted to better understand the role of dietary melatonin and melatonin precursors in bioavailable melatonin biological actions and their influence on oxidative status, besides the potential use of lentil sprouts and other melatonin-rich foods to combat oxidative stress-related diseases.

**Author Contributions:** Conceptualization, S.M.A. and M.A.M.-C.; Data curation, M.R.-H.; Formal analysis, M.R.-H., T.H., L.T.C., M.D. and P.R.-R.; Funding acquisition, M.A.M.-C.; Investigation, D.R.-C.; Methodology, Y.A., P.R.-R. and D.R.-C.; Supervision, Y.A., S.M.A. and M.A.M.-C.; Writing—original draft, M.R.-H; Writing—review & editing, M.R.-H, Y.A. and M.A.M.-C. All authors have read and agreed to the published version of the manuscript.

**Funding:** This research was funded by the Interuniversity Cooperation Projects from UAM-Banco Santander (2013/EEUU/02). M. Rebollo-Hernanz received funding from the FPU program of the Ministry of Science, Innovation, and Universities (Spain) for his predoctoral fellowship (FPU15/04238).

**Acknowledgments:** The authors acknowledge Dr. Juana Frias for her kind providing with the raw lentils.

**Conflicts of Interest:** The authors declare no conflict of interest.

## References

1. Arnao, M.B.; Hernández-Ruiz, J. The Potential of Phytomelatonin as a Nutraceutical. *Molecules* **2018**, *23*, 238. [CrossRef]
2. Loren, P.; Sánchez, R.; Arias, M.-E.; Felmer, R.; Risopatrón, J.; Cheuquemán, C. Melatonin Scavenger Properties against Oxidative and Nitrosative Stress: Impact on Gamete Handling and In Vitro Embryo Production in Humans and Other Mammals. *Int. J. Mol. Sci.* **2017**, *18*, 1119. [CrossRef]
3. Tan, D.-X.; Manchester, L.C.; Qin, L.; Reiter, R.J. Melatonin: A Mitochondrial Targeting Molecule Involving Mitochondrial Protection and Dynamics. *Int. J. Mol. Sci.* **2016**, *17*, 2124. [CrossRef] [PubMed]
4. Hardeland, R. Melatonin and the theories of aging: A critical appraisal of melatonin's role in antiaging mechanisms. *J. Pineal Res.* **2013**, *55*, 325–356. [CrossRef] [PubMed]
5. Zephy, D.; Ahmad, J. Type 2 diabetes mellitus: Role of melatonin and oxidative stress. *Diabetes Metab. Syndr. Clin. Res. Rev.* **2015**, *9*, 127–131. [CrossRef] [PubMed]
6. Cipolla-Neto, J.; Amaral, F.G.; Afeche, S.C.; Tan, D.X.; Reiter, R.J. Melatonin, energy metabolism, and obesity: A review. *J. Pineal Res.* **2014**, *56*, 371–381. [CrossRef]
7. Favero, G.; Franceschetti, L.; Buffoli, B.; Moghadasian, M.H.; Reiter, R.J.; Rodella, L.F.; Rezzani, R. Melatonin: Protection against age-related cardiac pathology. *Ageing Res. Rev.* **2017**, *35*, 336–349. [CrossRef]
8. Peuhkuri, K.; Sihvola, N.; Korpela, R. Dietary factors and fluctuating levels of melatonin. *Food Nutr. Res.* **2012**, *56*, 17252. [CrossRef]
9. Iriti, M.; Varoni, E.M. Melatonin in Mediterranean diet, a new perspective. *J. Sci. Food Agric.* **2015**, *95*, 2355–2359. [CrossRef]
10. Bonomini, F.; Borsani, E.; Favero, G.; Rodella, L.; Rezzani, R. Dietary Melatonin Supplementation Could Be a Promising Preventing/Therapeutic Approach for a Variety of Liver Diseases. *Nutr.* **2018**, *10*, 1135. [CrossRef]
11. Cardinali, D.P. Melatonin as a chronobiotic/cytoprotector: Its role in healthy aging. *Biol. Rhythm Res.* **2019**, *50*, 28–45. [CrossRef]
12. EFSA Panel on Dietetic Products, Nutrition and Allergies (NDA). Scientific Opinion on the substantiation of a health claim related to melatonin and reduction of sleep onset latency (ID **1698**, *1780*, 4080) pursuant to Article 13(1) of Regulation (EC) No 1924/2006. *EFSA J.* **2011**, *9*, 2241. [CrossRef]
13. Jiki, Z.; Lecour, S.; Nduhirabandi, F. Cardiovascular Benefits of Dietary Melatonin: A Myth or a Reality? *Front. Physiol.* **2018**, *9*, 528. [CrossRef] [PubMed]
14. Herrera, T.; Aguilera, Y.; Rebollo-Hernanz, M.; Bravo, E.; Benítez, V.; Martínez-Sáez, N.; Arribas, S.M.; del Castillo, M.D.; Martín-Cabrejas, M.A. Teas and herbal infusions as sources of melatonin and other bioactive non-nutrient components. *LWT* **2018**, *89*, 65–73. [CrossRef]
15. Meng, J.-F.; Shi, T.-C.; Song, S.; Zhang, Z.-W.; Fang, Y.-L. Melatonin in grapes and grape-related foodstuffs: A review. *Food Chem.* **2017**, *231*, 185–191. [CrossRef]
16. Tan, D.-X.; Zanghi, B.M.; Manchester, L.C.; Reiter, R.J. Melatonin identified in meats and other food stuffs: Potentially nutritional impact. *J. Pineal Res.* **2014**, *57*, 213–218. [CrossRef]
17. Aguilera, Y.; Rebollo-Hernanz, M.; Martín-Cabrejas, M.A. Melatonin. In *Legumes: Nutritional Quality, Processing and Potential Health Benefits*; Martín-Cabrejas, M.A., Ed.; CRC Press: Boca Raton, FL, USA, 2019; pp. 129–151. ISBN 9781482204056.

18. Martín-Cabrejas, M.A.; Aguilera, Y.; Benítez, V.; Reiter, R.J. Chapter 6—Melatonin Synthesis in Fermented Foods. In *Fermented Foods in Health and Disease Prevention*; Elsevier: New York, USA, 2017; pp. 105–129. ISBN 9780128023099.
19. Vaz Patto, M.C.; Amarowicz, R.; Aryee, A.N.A.; Boye, J.I.; Chung, H.-J.; Martín-Cabrejas, M.A.; Domoney, C. Achievements and Challenges in Improving the Nutritional Quality of Food Legumes. *CRC. Crit. Rev. Plant Sci.* **2015**, *34*, 105–143. [CrossRef]
20. Díaz, M.F.; Martínez, M.; Savón, L.L.; Torres, V.; Coto, G. CHAPTER 8. Obtaining, Chemically Characterizing and Nutritionally Evaluating Seasonal Legume Sprouts as a Feed Alternative. In *Legumes*; Royal Society of Chemistry: London, UK, 2019; pp. 177–195. [CrossRef]
21. Aguilera, Y.; Liébana, R.; Herrera, T.; Rebollo-Hernanz, M.; Sanchez-Puelles, C.; Benítez, V.; Martín-Cabrejas, M.A. Effect of illumination on the content of melatonin, phenolic compounds, and antioxidant activity during germination of lentils (*Lens culinaris* L.) and kidney beans (*Phaseolus vulgaris* L.). *J. Agric. Food Chem.* **2014**, *62*, 10736–10743. [CrossRef]
22. Kennaway, D.J. Are the proposed benefits of melatonin-rich foods too hard to swallow? *Crit. Rev. Food Sci. Nutr.* **2017**, *57*, 958–962. [CrossRef]
23. Bhattacharjee, P.; Probir, K.G.; Vernekar, M.; Singhal, R.S. *Role of Dietary Serotonin and Melatonin in Human Nutrition. Serotonin and Melatonin*; Taylor & Francis Group: Didcot, UK, 2016; pp. 317–332.
24. Abeysuriya, R.G.; Lockley, S.W.; Robinson, P.A.; Postnova, S. A unified model of melatonin, 6-sulfatoxymelatonin, and sleep dynamics. *J. Pineal Res.* **2018**, *64*, e12474. [CrossRef]
25. González-Flores, D.; Gamero, E.; Garrido, M.; Ramírez, R.; Moreno, D.; Delgado, J.; Valdés, E.; Barriga, C.; Rodríguez, A.B.; Paredes, S.D. Urinary 6-sulfatoxymelatonin and total antioxidant capacity increase after the intake of a grape juice cv. Tempranillo stabilized with HHP. *Food Funct.* **2012**, *3*, 34–39. [CrossRef] [PubMed]
26. Sae-Teaw, M.; Johns, J.; Johns, N.P.; Subongkot, S. Serum melatonin levels and antioxidant capacities after consumption of pineapple, orange, or banana by healthy male volunteers. *J. Pineal Res.* **2013**, *55*, 58–64. [CrossRef] [PubMed]
27. López-Martínez, L.X.; Leyva-López, N.; Gutiérrez-Grijalva, E.P.; Heredia, J.B. Effect of cooking and germination on bioactive compounds in pulses and their health benefits. *J. Funct. Foods* **2017**, *38*, 624–634. [CrossRef]
28. Ganesan, K.; Xu, B. Polyphenol-rich lentils and their health promoting effects. *Int. J. Mol. Sci.* **2017**, *11*, 2390. [CrossRef] [PubMed]
29. Aguilera, Y.; Herrera, T.; Liébana, R.; Rebollo-Hernanz, M.; Sanchez-Puelles, C.; Martín-Cabrejas, M.A. Impact of Melatonin Enrichment during Germination of Legumes on Bioactive Compounds and Antioxidant Activity. *J. Agric. Food Chem.* **2015**, *63*, 7967–7974. [CrossRef]
30. Zhang, Y.; Huo, M.; Zhou, J.; Xie, S. PKSolver: An add-in program for pharmacokinetic and pharmacodynamic data analysis in Microsoft Excel. *Comput. Methods Programs Biomed.* **2010**, *99*, 306–314. [CrossRef]
31. Singleton, V.L.; Orthofer, R.; Lamuela-Raventós, R.M. Analysis of total phenols and other oxidation substrates and antioxidants by means of Folin-Ciocalteu reagent. *Methods Enzymol.* **1998**, *299*, 152–178.
32. Arnao, M.B.; Hernández-Ruiz, J. Phytomelatonin, natural melatonin from plants as a novel dietary supplement: Sources, activities and world market. *J. Funct. Foods* **2018**, *48*, 37–42. [CrossRef]
33. Meng, X.; Li, Y.; Li, S.; Zhou, Y.; Gan, R.Y.; Xu, D.P.; Li, H.-B. Dietary sources and bioactivities of melatonin. *Nutr.* **2017**, *9*, 367. [CrossRef]
34. Rebollo-Hernanz, M.; Fernández-Gómez, B.; Herrero, M.; Aguilera, Y.; Martín-Cabrejas, M.A.; Uribarri, J.; del Castillo, M.D. Inhibition of the Maillard Reaction by Phytochemicals Composing an Aqueous Coffee Silverskin Extract via a Mixed Mechanism of Action. *Foods* **2019**, *8*, 438. [CrossRef] [PubMed]
35. Xu, M.; Jin, Z.; Peckrul, A.; Chen, B. Pulse seed germination improves antioxidative activity of phenolic compounds in stripped soybean oil-in-water emulsions. *Food Chem.* **2018**, *250*, 140–147. [CrossRef]
36. Di Stefano, E.; Tsopmo, A.; Oliviero, T.; Fogliano, V.; Udenigwe, C.C. Bioprocessing of common pulses changed seed microstructures, and improved dipeptidyl peptidase-IV and α-glucosidase inhibitory activities. *Sci. Rep.* **2019**, *9*, 15308. [CrossRef]
37. Xu, M.; Rao, J.; Chen, B. Phenolic compounds in germinated cereal and pulse seeds: Classification, transformation, and metabolic process. *Crit. Rev. Food Sci. Nutr.* **2019**, 1–20. [CrossRef] [PubMed]

38. Haytowitz, D.B.; Bhagwat, S. USDA Database for the Oxygen Radical Absorbance Capacity (ORAC) of Selected Foods, Release 2. *US Dep. Agric.* **2010**, *25*, 10–48.
39. Aguilera, Y.; Herrera, T.; Benítez, V.; Arribas, S.M.; López de Pablo, A.L.; Esteban, R.M.; Martín-Cabrejas, M.A. Estimation of scavenging capacity of melatonin and other antioxidants: Contribution and evaluation in germinated seeds. *Food Chem.* **2015**, *170*, 203–211. [CrossRef]
40. Yoshiara, L.; Madeira, T.; de Camargo, A.; Shahidi, F.; Ida, E. Multistep Optimization of β-Glucosidase Extraction from Germinated Soybeans (*Glycine max* L. Merril) and Recovery of Isoflavone Aglycones. *Foods* **2018**, *7*, 110. [CrossRef]
41. López, A.; El-Naggar, T.; Dueñas, M.; Ortega, T.; Estrella, I.; Hernández, T.; Gómez-Serranillos, M.P.; Palomino, O.M.; Carretero, M.E. Influence of Processing in the Phenolic Composition and Health-Promoting Properties of Lentils (*Lens culinaris* L.). *J. Food Process. Preserv.* **2017**, *41*, e13113. [CrossRef]
42. Aguilera, Y.; Rebollo-Hernanz, M.; Cañas, S.; Taladrid, D.; Martín-Cabrejas, M.A. Response surface methodology to optimise the heat-assisted aqueous extraction of phenolic compounds from coffee parchment and their comprehensive analysis. *Food Funct.* **2019**, *10*, 4739–4750. [CrossRef]
43. Harpsøe, N.G.; Andersen, L.P.H.; Gögenur, I.; Rosenberg, J. Clinical pharmacokinetics of melatonin: A systematic review. *Eur. J. Clin. Pharmacol.* **2015**, *71*, 901–909. [CrossRef]
44. Jana, S.; Rastogi, H. Effects of Caffeic Acid and Quercetin on In Vitro Permeability, Metabolism and In Vivo Pharmacokinetics of Melatonin in Rats: Potential for Herb-Drug Interaction. *Eur. J. Drug Metab. Pharmacokinet.* **2017**, *42*, 781–791. [CrossRef] [PubMed]
45. Hardeland, R. Taxon- and site-specific Melatonin catabolism. *Molecules* **2017**, *22*, 2015. [CrossRef] [PubMed]
46. Aghazadeh-Sanai, N.; Downs, J.L.; Mattison, J.A.; Ingram, D.K.; Kohama, S.G.; Urbanski, H.F. Effect of caloric restriction on plasma melatonin levels in aged rhesus macaques. *Neurobiol. Aging* **2017**, *55*, 213–216. [CrossRef] [PubMed]
47. Almeneessier, A.S.; Bahammam, A.S. How does diurnal intermittent fasting impact sleep, daytime sleepiness, and markers of the biological clock? Current insights. *Nat. Sci. Sleep* **2018**, *10*, 439–452. [CrossRef]
48. Delgado, J.; Pilar Terrón, M.; Garrido, M.; Pariente, J.A.; Barriga, C.; Rodríguez, A.B.; Paredes, S.D. A cherry nutraceutical modulates melatonin, serotonin, corticosterone, and total antioxidant capacity levels: Effect on ageing and chronotype. *J. Appl. Biomed.* **2012**, *10*, 109–117. [CrossRef]
49. Bravo, R.; Matito, S.; Cubero, J.; Paredes, S.D.; Franco, L.; Rivero, M.; Rodríguez, A.B.; Barriga, C. Tryptophan-enriched cereal intake improves nocturnal sleep, melatonin, serotonin, and total antioxidant capacity levels and mood in elderly humans. *Age* **2013**, *35*, 1277–1285. [CrossRef]
50. Khazaei, H.; Subedi, M.; Nickerson, M.; Martínez-Villaluenga, C.; Frias, J.; Vandenberg, A. Seed Protein of Lentils: Current Status, Progress, and Food Applications. *Foods* **2019**, *8*, 391. [CrossRef]
51. Reiter, R.J.; Manchester, L.C.; Tan, D.X. Melatonin in walnuts: Influence on levels of melatonin and total antioxidant capacity of blood. *Nutrition* **2005**, *21*, 920–924. [CrossRef]
52. Johns, N.P.; Johns, J.; Porasuphatana, S.; Plaimee, P.; Sae-Teaw, M. Dietary Intake of Melatonin from Tropical Fruit Altered Urinary Excretion of 6-Sulfatoxymelatonin in Healthy Volunteers. *J. Agric. Food Chem.* **2013**, *61*, 913–919. [CrossRef]
53. Maldonado, M.D.; Moreno, H.; Calvo, J.R. Melatonin present in beer contributes to increase the levels of melatonin and antioxidant capacity of the human serum. *Clin. Nutr.* **2009**, *28*, 188–191. [CrossRef]
54. McClements, D.J. Enhancing nutraceutical bioavailability through food matrix design. *Curr. Opin. Food Sci.* **2015**, *4*, 1–6. [CrossRef]
55. Hansen, M.B.; Witte, A.-B. The role of serotonin in intestinal luminal sensing and secretion. *Acta Physiol.* **2008**, *193*, 311–323. [CrossRef] [PubMed]
56. Martin, A.M.; Lumsden, A.L.; Young, R.L.; Jessup, C.F.; Spencer, N.J.; Keating, D.J. Regional differences in nutrient-induced secretion of gut serotonin. *Physiol. Rep.* **2017**, *5*, e13199. [CrossRef] [PubMed]
57. Lund, M.L.; Egerod, K.L.; Engelstoft, M.S.; Dmytriyeva, O.; Theodorsson, E.; Patel, B.A.; Schwartz, T.W. Enterochromaffin 5-HT cells – A major target for GLP-1 and gut microbial metabolites. *Mol. Metab.* **2018**, *11*, 70–83. [CrossRef]
58. Sumara, G.; Sumara, O.; Kim, J.K.; Karsenty, G. Gut-derived serotonin is a multifunctional determinant to fasting adaptation. *Cell Metab.* **2012**, *16*, 588–600. [CrossRef]

59. Zagajewski, J.; Drozdowicz, D.; Brzozowska, I.; Hubalewska-Mazgaj, M.; Stelmaszynska, T.; Laidler, P.M.; Brzozowski, T. Conversion L-tryptophan to melatonin in the gastrointestinal tract: The new high performance liquid chromatography method enabling simultaneous determination of six metabolites of L-tryptophan by native fluorescence and UV-VIS detection. *J. Physiol. Pharmacol.* **2012**, *63*, 613–621.
60. Fernstrom, J.D. A Perspective on the Safety of Supplemental Tryptophan Based on Its Metabolic Fates. *J. Nutr.* **2016**, *146*, 2601S–2608S. [CrossRef]
61. Zhao, H.; Wang, Y.; Yuan, B.; Liu, S.; Man, S.; Xu, H.; Lu, X. A novel LC-MS/MS assay for the simultaneous determination of melatonin and its two major metabolites, 6-hydroxymelatonin and 6-sulfatoxymelatonin in dog plasma: Application to a pharmacokinetic study. *J. Pharm. Biomed. Anal.* **2016**, *117*, 390–397. [CrossRef]
62. Tian, X.; Huo, X.; Dong, P.; Wu, B.; Wang, X.; Wang, C.; Liu, K.; Ma, X. Sulfation of melatonin: Enzymatic characterization, differences of organs, species and genders, and bioactivity variation. *Biochem. Pharmacol.* **2015**, *94*, 282–296. [CrossRef]
63. Oba, S.; Nakamura, K.; Sahashi, Y.; Hattori, A.; Nagata, C. Consumption of vegetables alters morning urinary 6-sulfatoxymelatonin concentration. *J. Pineal Res.* **2008**, *45*, 17–23. [CrossRef]
64. Howatson, G.; Bell, P.G.; Tallent, J.; Middleton, B.; McHugh, M.P.; Ellis, J. Effect of tart cherry juice (*Prunus cerasus*) on melatonin levels and enhanced sleep quality. *Eur. J. Nutr.* **2012**, *51*, 909–916. [CrossRef] [PubMed]
65. Cubero, J.; Valero, V.; Sánchez, J.; Rivero, M.; Parvez, H.; Rodríquez, A.B.; Barriga, C. The circadian rhythm of tryptophan in breast milk affects the rhythms of 6-sulfatoxymelatonin and sleep in newborn. *Neuroendocrinol. Lett.* **2005**, *26*, 657–661. [PubMed]
66. Huether, G.; Poeggeler, B.; Reimer, A.; George, A. Effect of tryptophan administration on circulating melatonin levels in chicks and rats: Evidence for stimulation of melatonin synthesis and release in the gastrointestinal tract. *Life Sci.* **1992**, *51*, 945–953. [CrossRef]
67. Huether, G. Melatonin Synthesis in the Gastrointestinal Tract and the Impact of Nutritional Factors on Circulating Melatonin. *Ann. N. Y. Acad. Sci.* **1994**, *719*, 146–158. [CrossRef] [PubMed]
68. Sánchez, S.; Sánchez, C.L.; Paredes, S.D.; Rodriguez, A.B.; Barriga, C. The effect of tryptophan administration on the circadian rhythms of melatonin in plasma and the pineal gland of rats. *J. Appl. Biomed.* **2008**, *6*, 177–186. [CrossRef]
69. Crozier, A.; Jaganath, I.B.; Clifford, M.N. Dietary phenolics: Chemistry, bioavailability and effects on health. *Nat. Prod. Rep.* **2009**, *26*, 1001–1043. [CrossRef]
70. Zhang, B.; Deng, Z.; Tang, Y.; Chen, P.X.; Liu, R.; Dan Ramdath, D.; Liu, Q.; Hernandez, M.; Tsao, R. Bioaccessibility, in vitro antioxidant and anti-inflammatory activities of phenolics in cooked green lentil (*Lens culinaris*). *J. Funct. Foods* **2017**, *32*, 248–255. [CrossRef]
71. Velderrain-Rodríguez, G.R.; Palafox-Carlos, H.; Wall-Medrano, A.; Ayala-Zavala, J.F.; Chen, C.Y.O.; Robles-Sánchez, M.; Astiazaran-García, H.; Alvarez-Parrilla, E.; González-Aguilar, G.A. Phenolic compounds: Their journey after intake. *Food Funct.* **2014**, *5*, 189–197. [CrossRef]
72. Domínguez-Avila, J.A.; Wall-Medrano, A.; Velderrain-Rodríguez, G.R.; Chen, C.Y.O.; Salazar-López, N.J.; Robles-Sánchez, M.; González-Aguilar, G.A. Gastrointestinal interactions, absorption, splanchnic metabolism and pharmacokinetics of orally ingested phenolic compounds. *Food Funct.* **2017**, *8*, 15–38. [CrossRef]
73. Hasslauer, I.; Oehme, A.; Locher, S.; Valotis, A.; Van't Slot, G.; Humpf, H.-U.; Schreier, P. Flavan-3-ol C-glycosides—Preparation and model experiments mimicking their human intestinal transit. *Mol. Nutr. Food Res.* **2010**, *54*, 1546–1555. [CrossRef]
74. Garrido, M.; Paredes, S.D.; Cubero, J.; Lozano, M.; Toribio-Delgado, A.F.; Munoz, J.L.; Reiter, R.J.; Barriga, C.; Rodriguez, A.B. Jerte Valley Cherry-Enriched Diets Improve Nocturnal Rest and Increase 6-Sulfatoxymelatonin and Total Antioxidant Capacity in the Urine of Middle-Aged and Elderly Humans. *J. Gerontol. A Biol. Sci. Med. Sci.* **2010**, *65*, 909–914. [CrossRef] [PubMed]
75. Selby-Pham, S.N.B.; Cottrell, J.J.; Dunshea, F.R.; Ng, K.; Bennett, L.E.; Howell, K.S. Dietary phytochemicals promote health by enhancing antioxidant defence in a pig model. *Nutr.* **2017**, *9*, 758. [CrossRef] [PubMed]
76. Reiter, R.J.; Mayo, J.C.; Tan, D.-X.; Sainz, R.M.; Alatorre-Jimenez, M.; Qin, L. Melatonin as an antioxidant: Under promises but over delivers. *J. Pineal Res.* **2016**, *61*, 253–278. [CrossRef] [PubMed]
77. Salehi, B.; Sharopov, F.; Fokou, P.V.T.; Kobylinska, A.; Jonge, L.; Tadio, K.; Sharifi-Rad, J.; Posmyk, M.M.; Martorell, M.; Martins, N.; et al. Melatonin in Medicinal and Food Plants: Occurrence, Bioavailability, and Health Potential for Humans. *Cells* **2019**, *8*, 681. [CrossRef] [PubMed]

78. Szewczyk-Golec, K.; Rajewski, P.; Gackowski, M.; Mila-Kierzenkowska, C.; Wesołowski, R.; Sutkowy, P.; Pawłowska, M.; Woźniak, A. Melatonin Supplementation Lowers Oxidative Stress and Regulates Adipokines in Obese Patients on a Calorie-Restricted Diet. *Oxid. Med. Cell. Longev.* **2017**, *2017*, 8494107. [CrossRef]
79. Parkar, S.G.; Kalsbeek, A.; Cheeseman, J.F. Potential role for the gut microbiota in modulating host circadian rhythms and metabolic health. *Microorganisms* **2019**, *7*, 41. [CrossRef] [PubMed]

*Article*

# Sprouted Barley Flour as a Nutritious and Functional Ingredient

**Daniel Rico [1], Elena Peñas [2,*], María del Carmen García [1], Cristina Martínez-Villaluenga [2], Dilip K. Rai [3], Rares I. Birsan [3], Juana Frias [2] and Ana B. Martín-Diana [1]**

[1] Subdirection of Research and Technology, Agro-Technological Institute of Castilla y León, Consejería de Agricultura y Ganadería, Finca de Zamadueñas, 47171 Valladolid, Spain; ricbarda@itacyl.es (D.R.); gargurmar@itacyl.es (M.d.C.G.); mardiaan@itacyl.es (A.B.M.-D.)

[2] Department of Food Characterization, Quality and Safety, Institute of Food Science, Technology and Nutrition (ICTAN-CSIC), 28006 Madrid, Spain; c.m.villaluenga@csic.es (C.M.-V.); frias@ictan.csic.es (J.F.)

[3] Department of Food BioSciences, Teagasc Food Research Centre Ashtown, 15 Dublin, Ireland; Dilip.Rai@teagasc.ie (D.K.R.); Rares.Birsan@teagasc.ie (R.I.B.)

* Correspondence: elenape@ictan.csic.es; Tel.: +34-91-258-76

Received: 10 February 2020; Accepted: 2 March 2020; Published: 5 March 2020

**Abstract:** The increasing demand for healthy food products has promoted the use of germinated seeds to produce functional flours. In this study, germination conditions were optimized in barley grains with the aim to produce flours with high nutritional and biofunctional potential using response surface methodology (RSM). The impact of germination time (0.8–6 days) and temperature (12–20 °C) on barley quality was studied. Non-germinated barley was used as the control. The content of vitamins $B_1$, $B_2$ and C, and proteins increased notably after germination, especially at longer times, while levels of fat, carbohydrates, fibre, and β-glucan were reduced. Total phenolic compounds, γ-aminobutyric acid and antioxidant activity determined by Oxygen Radical Absorbance Capacity increased between 2-fold and 4-fold during sprouting, depending on germination conditions and this increase was more pronounced at higher temperatures (16–20 °C) and longer times (5–6 days). Procyanidin B and ferulic acid were the main phenolics in the soluble and insoluble fraction, respectively. Procyanidin B levels decreased while bound ferulic acid content increased during germination. Germinated barley flours exhibited lower brightness and a higher glycemic index than the control ones. This study shows that germination at 16 °C for 3.5 days was the optimum process to obtain nutritious and functional barley flours. Under these conditions, sprouts retained 87% of the initial β-glucan content, and exhibited levels of ascorbic acid, riboflavin, phenolic compounds and GABA between 1.4-fold and 2.5-fold higher than the non-sprouted grain.

**Keywords:** barley; germination; flour; RSM; nutritional properties; bioactive compounds; quality

## 1. Introduction

Nowadays, the main changes in demand for agricultural and food products are being fueled by population growth and lifestyle modifications. Many food manufacturers today are looking to replace wheat flour by alternative flours to be included in new formulations -with high nutritive value and bioactive properties.

Barley (*Hordeum vulgare* L.) is an ancient cereal which has been traditionally used for animal feeding and as raw material in malting industry [1]. Today, barley remains an important grain in some cultures of Asia and Northern Africa, and in recent years, there has been an increasing interest in using barley as an ingredient in food products due its nutritional value and high content of biologically active compounds [2]. This renewed trend in formulating novel barley foods is based on the beneficial effects of β-glucans on reducing blood cholesterol and glycemic index. In fact, the European Commission has authorized health claims linking barley β-glucan with a reduction of blood cholesterol. This claim can be used in foods with β-glucan levels ≥1 g [3]. The high β-glucan content in barley grains makes it an attractive ingredient for development of novel functional flours [4]. In addition to β-glucan, barley contains some other bioactive compounds (phenolic compounds, tocols and sterols) with beneficial effects that are currently under extensive investigation [5]. Therefore, there is a large potential to include barley flour partially or as a whole grain in a wide range of cereal-based food products as an alternative to commonly used cereals (wheat, rice, maize and oat).

Most barley varieties are hulled, although several hull-less varieties are also cultivated. Hull-less barley, known as 'naked' barley, requires minimal cleaning compared to hulled barley, as no processing is needed to remove the inedible outer hull [6]. The use of hull-less barley varieties with high levels of β-glucans is interesting to obtain functional barley flours that can be easily incorporated into foods to meet the authorized health claims of β-glucan. Barley flours usually have a higher water absorption ability than wheat flours for its greater content of soluble fiber [7], which can positively affect their baking properties.

Nowadays, there is a strong demand for food products with health-promoting properties. This is reflected by the growing market of foods that contain functional ingredients able to modulate different body functions, improving the health of consumers [8]. The scientific community is focusing its efforts on identifying novel functional ingredients that will expand the options to develop innovative food products with health benefits. Barley flours enriched in bioactive compounds are attractive ingredients to achieve these purposes, since they can be easily incorporated into different food matrices. Germination can be explored as a low-cost process to enhance the levels of functional compounds with healthy attributes in barley. Germination has been demonstrated to be an inexpensive and sustainable process that improves the nutritive quality and the content of functional compounds of grains, but also their palatability, digestibility and bioavailability [9,10]. However, the magnitude of changes caused by germination depends on the grain variety and germination conditions. For this reason, the optimization of germination parameters is crucial for improving the nutritional and bioactive properties of a selected grain variety. Although different studies have investigated the influence of germination in several cereals and legumes, few studies have optimized germination conditions in barley for producing high-quality flours. Thus, the objective of this study was to explore the effect of germination temperature and time on the nutritional, bioactive and other quality features in a hull-less barley variety. The functional sprouted flour obtained could be used for the production of functional bakery products (bread, cookies, cakes, muffins, etc).

## 2. Materials and Methods

### 2.1. Chemicals

α-Amylase from porcine pancreas (EC 3.2.1.1), 2,2'-Azinobis 3-ethylbenzothiazoline-6-sulfonic acid (ABTS), 2,2'-diazobis-(2-aminodinopropane)-dihydrochloride (AAPH), 2,2-diphenyl-1-picrylhydrazyl (DPPH), Folin–Ciocalteu (FC) reagent, gallic acid (GA), 6-hydroxy-2,5,7,8-tetramethyl-2-carboxylic acid (Trolox), and sodium carboxymethylcellulose (CMC) were obtained from Sigma-Aldrich, Co. (St. Louis, MO, USA). Amyloglucosidase (EC 3.2.1.3), glucose oxidase-peroxidase (GOPOD) and 1.3:1.4 mixed-linkage β-glucan kits were provided by Megazyme International Ireland (Wicklow, Ireland).

### 2.2. Barley Grains

A hull-less H13 barley variety obtained by conventional breeding between Merlin and Volga varieties in Agro-Technological Institute of Castilla y León was used. This barley variety, commercially known as GALIS, contains high levels of β-glucan (5.2 g/100 g). Whole-grain flour was obtained in order to compare the results obtained in sprouted grains with those of raw grain andevaluate the changes in nutrients, bioactive compounds and other quality parameters caused by the germination process. For these purposes, whole grains were milled by using a mechanical grinder (Moulinex,

France), sifted through a 60-mesh sieve (0.5 mm pore diameter) and stored under vacuum at −20 °C. This flour was also analyzed.

### 2.3. Germination

Twenty grams of barley seeds was soaked in 0.1% sodium hypochlorite at 1:5 ratio (*w*/*v*) for 30 min and then washed with tap water until it reached a neutral pH. Then, seeds were steeped in tap water (at the same ratio) at room temperature for 4 h. Water was removed and seeds were spread across moist filter over steel grid, which was placed in plastic trays containing sterile tap water. Seeds were covered by a filter paper and introduced in a germination cabinet (model MLR-350, Sanyo, Japan), with >90% air humidity. Germination was carried out in darkness at temperatures in the range 12.1–19.9 °C and times of 1.6–6.19 days, conditions chosen according to the response surface model (Table 1). The germination process was carried out in triplicate for each germination condition. The sprouted seeds were freeze-dried (LyoQuest, Telstar, Spain), milled (Moulinex, France), sifted through a 60-mesh sieve (0.5 mm pore diameter) and stored under vacuum at −20 °C until further analysis.

**Table 1.** Independent variables (factors) of the response surface central composite design.

| Factor A<br>Temperature (°C) | Factor B<br>Time (Days) | Experiments<br>(Codes) |
|---|---|---|
| 12.1 | 3.58 | 1 |
| 13.2 | 1.60 | 2 |
| 13.2 | 5.52 | 3 |
| 16.0 | 0.81 | 4 |
| 16.0 | 3.48 | 5* |
| 16.0 | 6.19 | 6 |
| 18.8 | 1.60 | 7 |
| 18.8 | 5.52 | 8 |
| 19.9 | 3.60 | 9 |

*Center point. Three replicate experiments were performed for the center point.

### 2.4. Experimental Design

The response surface methodology (RSM) was employed to evaluate the impact of germination time and temperature on different quality and bioactive properties of sprouted barley. A central composite design was used with a total of 11 experimental assays, including three center points and "star" points to estimate curvature. The order of the experiments was randomly selected to avoid the effects of lurking variables (Table 1 and Figure 1). The combination of the two factors (germination temperature and time) studied in the response surface design and its optimization was based on the +1 and −1 variable levels. Proximal, nutritional, bioactive and sensory parameters were selected as dependent variables. The design was used to explore the quadratic response surfaces and to obtain polynomial equations for each response variable. The optimization of germination conditions was performed through the desirability function approach that allows identifying the most suitable combination of germination time and temperature to achieve the maximum desirable response values.

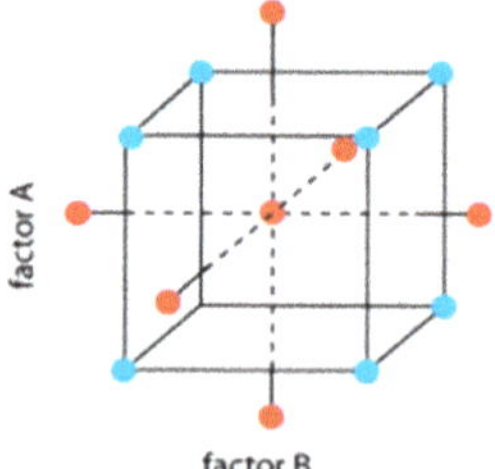


**Figure 1.** Response surface central composite design with two independent variables (factors), three center points and two star points.

## 2.5. Proximate and Fatty Acid Compositions

The moisture contents of raw and germinated flours obtained from barley were determined by drying 5 ± 0.001 g sample flour at 105 °C for 3 h. Total protein content was assessed by the Dumas method. A factor of 6.25 was used to transform nitrogen into protein values. Total fat content was evaluated after extraction of sample flour with petroleum ether (BP 40–60 °C) during 4 h in a Soxtec fat-extracting unit. Carbohydrates were estimated by difference. Total dietary fiber content was determined by a commercial kit provided (Sigma, St. Louis, MO, USA), by following the manufacturer's instructions. Ash content was evaluated by incineration of samples in a muffle furnace at 550 °C for 5 h. The results were expressed as g/100 g of dry matter (d.m.), after correcting them for moisture content.

For the evaluation of fatty acid composition in sample flours, the lipid-containing chloroform phase was separated and evaporated. Then, the sample was dissolved in 1 mL of hexane and methylated by the addition of 0.5 M methanolic KOH (100 µL) and incubation for 10 min at room temperature. An analysis of fatty acid methyl esters (FAME) in the upper layer obtained was carried out on a gas chromatograph (Agilent 7890A, Agilent Technologies, CA, USA) coupled with a flame ionization detector.

## 2.6. Content of Vitamin C

Vitamin C content was determined in barley flour samples by homogenization of the sample with 50 mL of oxalic acid (2 g/100 g) and titration with 2,6,-dichlorolindophenol (0,25 g/L) using a potentiometric titrator (877 Tritino Plus, Metrohm, Switzerland). The results were expressed as mg vitamin/100 g d.m.

## 2.7. Content of Thiamine and Riboflavin

Both vitamins were extracted from raw and germinated barley flour by acid and enzymatic hydrolysis and quantified by reversed-phase high-performance liquid chromatography (HPLC) coupled to a fluorescence detector, according to Frias et al. [11]. The results were expressed as µg vitamin/100 g d.m.

## 2.8. Content of β-Glucan

The β-glucan content was quantified by 1.3:1.4 mixed-linkage β-glucan kit (Megazyme, Ireland), following the manufacturer instructions. The assay uses lichenase and β-glucosidase to hydrolyse β-glucan to glucose. Subsequently, glucose reacts with GOPOD (glucoseoxidase/peroxidase) reagent and the absorbance at 510 nm of the resultant compound was measured in a microplate reader Synergy HT (BioTek Instruments, Winooski, VT, USA). All measurements were performed in triplicate. The results were expressed as g β-glucan/100 g d.m.

### 2.9. Content of γ-Aminobutyric Acid (GABA)

GABA was quantified in water extracts obtained from raw and germinated barley samples using HPLC coupled to a diode array detector as previously described [9]. Results were expressed as mg/100 g d.m.

### 2.10. Total Phenolic Content Determined by the Folin–Ciocalteu Method

Total phenolic contents (TPCs) were evaluated in methanolic extracts obtained from control and sprouted barley flours. Briefly, 1 g of sample was extracted with 10 mL of methanol:water (1:1, *v/v*; acidified to pH 2 with 0.1 M HCl) by stirring (250 rpm, 25 °C, 1 h) in an orbital shaker. Samples were centrifuged (25 °C, 3800× *g*, 10 min), and the supernatant was collected, filtered (Whatman paper n° 1), and made into a final volume of 25 mL with the extracting solvent. Extracts were stored at −80 °C until use. TPCs were estimated using the Folin–Ciocalteu method as previously reported [12]. Gallic acid (GA) was used as the standard. The final results were corrected for moisture and expressed as μmol GA equivalents (GA Eq)/g d.m.

### 2.11. Content and Profile of Free and Bound Phenolic Compounds

Free phenolic compounds were extracted as reported above (see Section 2.9). The residues obtained following the free phenolic extraction were subjected to alkaline and acid hydrolysis for recovering the bound (insoluble) fraction, as described by Mattila et al. [13]. Soluble and insoluble phenolic extracts were filtered through a 0.22-μm filter, lyophilized and stored at −20 °C until analysis.

The profile of free and bound phenolic compounds in barley samples was analyzed by HPLC coupled with a mass spectrometer, according to Birsan et al. [14]. The results were expressed as μg/mg of dry extract.

### 2.12. Total Antioxidant Activity

Total antioxidant activity was measured in methanolic extracts previously obtained using classical assays, namely DPPH radical scavenging activity, Oxygen Radical Absorbance Capacity (ORAC), Ferric Reducing Ability of Plasma (FRAP) and Trolox Equivalent Antioxidant Capacity (TEAC). Moreover, the DPPH method without any extraction procedure (Q-DPPH) was used to evaluate the total antioxidant activity of the samples.

#### 2.12.1. DPPH Assays (Classical and Q-Versions)

The classical version of DPPH assay was performed as previously described [15], with slight modifications. An amount of 25 μL of extracts was mixed with 100 μL of MilliQ water and 125 μL of DPPH· working solution (100 μM using methanol as solvent) in a 96-well microplate. Absorbance at 515 nm was recorded for 30 min in a microplate reader.

The Q-DPPH assay was carried out by mixing 10 mg of barley samples with 1.6 mL DPPH· working solution (50 μM) prepared in either methanol (Q-DPPH_M) or methanol:water (50:50, *v:v*) (Q-DPPH_MW). A volume of 160 μL of methanol:water (50:50, *v:v*; pH = 2) was added for the calibration curve with Trolox to compensate for the solvent present in the sample reactions. The samples were incubated in an orbital shaker (750 rpm for 30 min) then centrifuged (10,000× *g* for 2 min). Absorbance was recorded at 515 nm in a microplate reader. The results were corrected for moisture and expressed as μg Trolox equivalents/100 g d.m.

#### 2.12.2. TEAC Assay

The classical version of this method was carried out according to Re et al. [16] with slight modifications. An aqueous ABTS solution (7 mM) was mixed with 2.45 mM $K_2O_8S_2$ in a 1:1 (*v/v*) ratio (stock $ABTS^{\cdot+}$ solution). Stock $ABTS^{\cdot+}$ solution was diluted with phosphate buffer (75 mM, pH = 7.4) ($ABTS^{\cdot+}$ working solution), with an absorbance value of 0.70 ± 0.02 at 734 nm. Then, 20 μL of diluted samples was mixed with 200 μL ABTS·$^{+}$ working solution in a 96-well microplate. The decay in absorbance at 730 nm was recorded over 30 min with a microplate reader. Trolox was used as the standard. The results were corrected for moisture and expressed μg Trolox equivalents/100 g d.m.

#### 2.12.3. ORAC Assay

ORAC assay was carried out as earlier described [17] with some modifications. Trolox standard curve (15-240 mM) and samples were diluted in phosphate buffer (10 mM, pH 7.4). Fluorescein (150 μL) was placed in a 96-well black microplate and then 25 μL of Trolox standard, sample or phosphate buffer (blank) was added. After incubation (37 °C, 3 min), AAPH solution was added to initiate the oxidation reaction. Fluorescence was monitored over 35 min with a microplate reader at excitation and emission wavelengths of 485 nm and 528 nm, respectively. The results were expressed as mmol Trolox Equivalents (TE)/g d.m.

#### 2.12.4. FRAP Assay

The assay was performed following the protocol reported by Vijayalakshmi and Ruckmani [18]. Results were expressed as mmol Fe equivalents/g d.m.

### 2.13. *Glycemic Index*

For the determination of glycemic index (GI) in sprouted barley, firstly, the content of available starch was measured using the total starch assay kit of Megazyme (K-TSTA 08/16). Afterwards, the in vitro starch hydrolysis rate was determined as described by Gularte and Rosell [19], with slight modifications. Samples containing 50 mg of available starch were dissolved in Tris-maleate buffer (0.1 M, pH = 6, 2 mL) and then 2 mL enzymatic solution containing porcine pancreatic α-amylase (460 U/mL) and amyloglucosidase (6.6 U/mL) were added. Aliquots were taken during the incubation period (150 min), were kept in boiling water for 5 min to stop the enzymatic reaction and cooled in ice. Then, a volume of 150 μL of absolute ethanol was added and the sample was centrifuged (4 °C, 10,000× *g* for 5 min). The pellet was washed with 150 μL ethanol:water (1:1, *v*/*v*) and the supernatants were pooled together and stored at 4 °C for the subsequent colorimetric analysis of reducing sugars using the GOPOD kit (Megazyme, Bray, Ireland). GI values were calculated from hydrolysis index (HI) values, as proposed by Granfeldt [20].

### 2.14. *Colorimetric Characterizationof Germinated Barley Flours*

Color was measured using a colorimeter (Minolta CM-2002, Osaka, Japan) with D65 as illuminant and 45/0 sensor. The instrument was calibrated with a white tile standard (L*593.97, a*520.88 and b*51.21). The CIE L*a*b* parameters were converted to Hue (arctan b*/a*), Chroma $(a^{*2} + b^{*2})^{1/2}$ based in the CIELab (L*, a* and b*).

### 2.15. *Sensory Analysis*

Panelists (20–45 years old) with previous experience in sensory analysis were recruited from the Staff of the Agro-Technological Institute of Castilla y León. Panelists were instructed to rinse their mouths with water between samples. Samples were subjected to descriptive test. Colour, texture and flavour were evaluated for all the samples using a scale from 0 (disliked extremely) to 5 (liked extremely).

### 2.16. *Statistical Analysis*

Analytical determinations were carried out in duplicate for each replicate (three replicates were obtained for each germination condition). The equations obtained by RSM models were solved using the least squares (LS) method. Analysis of variance (ANOVA) was used to assess the reliability of the regression equations obtained and to determine the significance of all terms (linear, quadratic and interaction) of each factor at a probability level ($p$) of 0.05. The goodness of models to describe the variations of experimental data was also quantified by the coefficient of determination ($R^2$). A quadratic model was used to evaluate the responses. A statistical analysis was performed using Statgraphic Centurion XVI (Rockville, MD, USA) software.

## 3. Results

The nutritional, bioactive and sensory properties of germinated barley were evaluated through different response variables using polynomic regression models (Table 2). ANOVA results reveal that the predictive models obtained for all response variables were statistically significant ($p \leq 0.05$), with no significant values of lack-of-fit, indicating that the models satisfactorily predict the relationship between response variables and independent factors. Higher $R^2$ values than 0.75 suggest an adequate fitting of the polynomial models to the experimental values. The germination yield was higher than 50% regardless of the conditions used during sprouting.

**Table 2.** Polynomic equations and optimum values for the dependent variables studied in the RSM design.

| Response Variables | Model Equations | Optimum Temperature (Tp) | Optimum Time (Tm) |
|---|---|---|---|
| Fat | $4.984 - 0.214 \times TP + 0.022 \times TM + 0.006 \times TP^2 - 0.008 \times TM \times TP - 0.009 \times TP^2$ | 12.1 | 0.81 |
| Protein | $19.066 - 0.551 \times TM + 0.167 \times TP + 0.019 \times TM^2 + 0.010 \times TM \times TP - 0.026 \times TP^2$ | 19.9 | 6.19 |
| Fiber | $35.541 - 2.362 \times TP - 1.176 \times TM + 0.080 \times TP^2 + 0.00004 \times TM \times TP + 0.098 \times TP^2$ | 19.9 | 0.81 |
| Carbohydrates | $65.948 - 0.075 \times TM + 2.804 \times TP + 0.025 \times TM^2 - 0.098 \times TM \times TP - 0.121 \times TP^2$ | 12.1 | 0.81 |
| Saturated fatty acids | $24.444 - 0.587 \times TM - 2.426 \times TP + 0.019 \times TM^2 + 0.179 \times TP \times TM + 0.010 \times TP^2$ | 19.9 | 6.19 |
| Unsaturated fatty acids | $75.556 + 0.587 \times TM + 2.426 \times TP - 0.019 \times TM^2 - 0.179 \times TM \times TP - 0.100 \times TP^2$ | 12.1 | 0.81 |
| Monounsat. fatty acids | $17.847 + 0.398 \times TM + 0.288 \times TP - 0.010 \times TP^2 - 0.123 \times TM \times TP + 0.036 \times TP^2$ | 14.3 | 0.81 |
| Polyunsat. fatty acids | $57.747 + 0.184 \times TM + 2.141 \times TP - 0.008 \times TP^2 - 0.056 \times TM \times TP - 0.136 \times TP^2$ | 12.1 | 5.4 |
| Vitamin C | $61.315 - 6.418 \times TP + 0.101 \times TM + 0.110 \times TP^2 + 0.060 \times TM \times TP - 0.020 \times TP^2$ | 19.9 | 6.19 |
| Vitamin $B_1$ | $1003.83 - 30.743 \times TM - 37.626 \times TP + 0.867 \times TM^2 + 0.059 \times TP \times TM + 4.207 \times TM^2$ | 12.1 | 0.81 |
| Vitamin $B_2$ | $4.880 - 0.329 \times TP + 39.175 \times TM + 0.290 \times TP^2 - 0.460 \times TM \times TP - 1.271 \times TP^2$ | 19.9 | 6.19 |
| β-glucan | $6.886 + 0.052 \times TP - 1.68 \times TM - 0.009 \times TP^2 + 0.082 \times TP \times TM + 0.014 \times TP^2$ | 12.1 | 0.81 |
| GABA | $311.39 - 28.559 \times TM - 4.590 \times TP + 0.759 \times TM^2 + 2.101 \times TP \times TM - 1.924 \times TM^2$ | 19.9 | 6.19 |
| TPC | $367.582 - 31.560 \times TP - 8.571 \times TM + 0.874 \times TM^2 + 1.297 \times TP \times TM - 0.929 \times TP^2$ | 19.9 | 6.19 |
| Bound ferulic acid | $423.975 - 36.487 \times TP - 47.023 \times TM + 0.784 \times TP^2 + 2.769 \times TM \times TP + 0.212 \times TP^2$ | 12.1 | 0.81 |
| Free procyanidin | $53.593 - 5.589 \times TP - 2.911 \times TM + 0.147 \times TP^2 + 0.181 \times TM \times TP - 0.025 \times TP^2$ | 12.1 | 0.81 |

**Table 2.** *Cont.*

| Response Variables | Model Equations | Optimum Temperature (Tp) | Optimum Time (Tm) |
|---|---|---|---|
| FRAP | $0.407 - 0.027 \times TP - 0.017 \times TM + 0.0007 \times TP^2 + 0.0009 \times TM \times TP + 0.0005 \times TP^2$ | 19.9 | 6.19 |
| ORAC | $205.967 - 23.180 \times TM - 0.708 \times TP + 0.709 \times TM^2 + 0.923 \times TM \times TP - 1.361 \times TP^2$ | 19.9 | 6.19 |
| DPPH | $4274.09 - 291.532 \times TP - 284.032 \times TM + 5.985 \times TP^2 + 18.461 \times TM \times TP - 0.231 \times TP^2$ | 12.1 | 0.81 |
| TEAC | $15296.3 - 1782.21 \times TM + 1758.61 \times TP + 59.311 \times TM^2 - 21.386 \times TP \times TM - 119.929 \times TP^2$ | 19.9 | 6.19 |
| DPPH-Q * | $5252.4 - 507.735 \times TP - 136.6 \times TM + 14.287 \times TP^2 + 11.670 \times TM \times TP - 0.503 \times TM^2$ | 19.9 | 6.19 |
| L* | $100.793 - 1.859 \times TM + 0.478 \times TP + 0.054 \times TM^2 - 0.003 \times TM \times TP - 0.069 \times TP^2$ | 12.1 | 3.2 |
| a* | $0.455 + 0.088 \times TM + 0.182 \times TP - 0.002 \times TM^2 - 0.013 \times TM \times TP - 0.0004 \times TP^2$ | 19.9 | 0.81 |
| b* | $16.906 - 0.874 \times TM - 0.564 \times TP + 0.025 \times TM^2 + 0.057 \times TP \times TM - 0.014 \times TP^2$ | 12.1 | 0.81 |
| Glycemic index | $105.067 - 4.918 \times TM - 3.373 \times TP + 0.180 \times TM^2 + 0.093 \times TP \times TM + 0.471 \times TP^2$ | 19.9 | 6.19 |
| Color | $(-) 0.322 + 0.0469 \times TM + 0.075 \times TP$ | 12.1 | 0.81 |
| Texture | $0.268 + 0.039 \times TM - 0.011 \times TP$ | 12.1 | 0.81 |
| Flavour | $(-) 0.365 + 0.047 \times TM + 0.085 \times TP$ | 19.9 | 6.19 |

FRAP: Ferric Reducing Ability of Plasma; ORAC: Oxygen Radical Absorbance Capacity; DPPH: 2,2-difenil-1-picrylhydrazyl Radical Scavenging Activity; TEAC:Trolox Equivalent Antioxidant Capacity; DPPH-Q: Direct DPPH method (without extraction); L*: Luminosity; a*: CIELAB a* parameter; b*: CIELAB b* parameter.

### 3.1. *Effect of Germination on Proximate Composition of Sprouted Barley Flours*

The polynomic regression equations (Table 2) and response surface tridimensional contour plots (Figure 2) obtained for proximate composition of germinated barley showed a significant higher effect of germination time than temperature on these nutritional parameters.

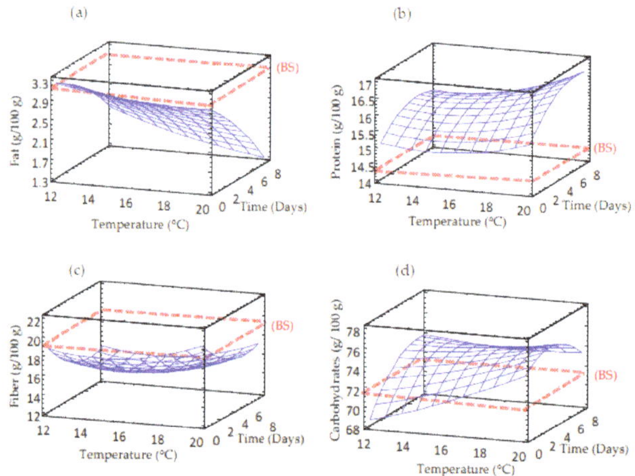


**Figure 2.** Response surface plots for proximal composition in germinated barley flour as a function of two independent variables (time and temperature). (**a**) Fat content; (**b**) Protein content; (**c**) Fiber content; (**d**) Carbohydrate content. BS: barley seed flour (non-germinated).

Fat levels of barley sprouts were significantly reduced compared to non-germinated grain and this decrease showed a linear time-dependent behavior (Figure 2a). Decrease of fat may be linked to depletion of the stored fat that is involved in grain catabolic activities required for protein synthesis in the growing plant [21].

Contrary results were obtained for protein content, which increased significantly after germination in all barley sprouts regardless of the germination conditions used (Figure 2b). The enhancement of protein content in seed sprouts might be attributed to the loss of carbohydrates during respiration that cause an apparent increase in other nutrients such as proteins, as has been previously reported in chia sprouts [22]. Higher temperature and longer time would produce greater loss in the sprouts dry weight and a more pronounced enhancement of protein levels. Moreover, there is a reawakening protein synthesis upon imbibition [23], which could also partially explain the increased protein content in sprouted barley. Again, the protein increase was only dependent on germination time, where a linear effect was observed.

The content of fiber in germinated barley decreased with respect to non-germinated grains (Figure 2c), that could be associated with the activation of hydrolytic enzymes, mainly at longer periods of germination. However, no significant differences were observed at different germination temperatures.

The carbohydrate content of sprouted barley decreased at low temperatures and short germination times and remained almost constant at the other temperatures (Figure 2d).

### 3.2. *Effect of Germination on Fatty Acid Content of Sprouted Barley Flours*

The content of saturated, unsaturated, monounsaturated and polyunsaturated fatty acids was also evaluated in germinated barley (Figure 3 and Table 2). Barley germinated at lower temperatures and shorter times exhibited levels of these compounds similar to non-germinated barley grain. Saturated fatty acids duplicated during sprouting, reaching values close to 37% at longer times of germination (Figure 3a) and the opposite behavior was observed for unsaturated fatty acids (Figure 3b). Polyunsaturated acids (Figure 3d) did not show significant differences compared to control barley. The values ranged from 59% to 63% with an insignificant decrease at higher temperatures and longer times of germination. Our results are in line with previous investigations showing that triglycerides that represent the major lipids storage form in cereals, are hydrolyzed by lipases to diglycerides, monoglycerides, and then to glycerol and fatty acids [24]. Hence, it seems that α-oxidation plays a minor role in sprouting grains while β-oxidation plays a key role with the aid of β-oxidase, yielding energy for metabolic pathways [25].

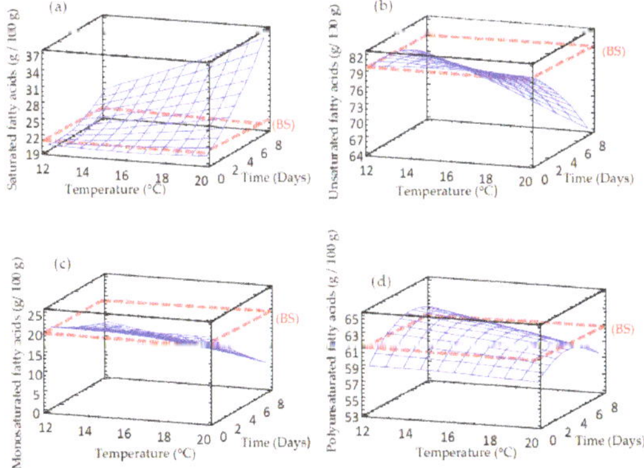


**Figure 3.** Response surface plots for fatty acid content in germinated barley flour as a function of two independent variables (time and temperature). (**a**) Saturated fatty acids; (**b**) Unsaturated fatty acids; (**c**) Monounsaturated fatty acids; (**d**) Polyunsaturated fatty acids. BS: barley seed flour (non-germinated).

### 3.3. Effect of Germination on Vitamin Content of Sprouted Barley Flours

Vitamin content increased in barley after germination and it exhibited a significant linear dependence respect to temperature and time regardless of the vitamin studied. Higher levels of water-soluble vitamins ($B_1$, $B_2$ and C) were observed at longer germination times and higher temperatures (Figure 4). Vitamins $B_1$ (Figure 4a) and C (Figure 4c) contents almost triplicated in barley germinated at longer time while vitamin $B_2$ (Figure 4b) showed a six-fold increase in sprouted barley. Different authors have reported an increase of vitamins content during chickpea and barley germination as consequence of the biosynthesis undergone in the grain [26,27].

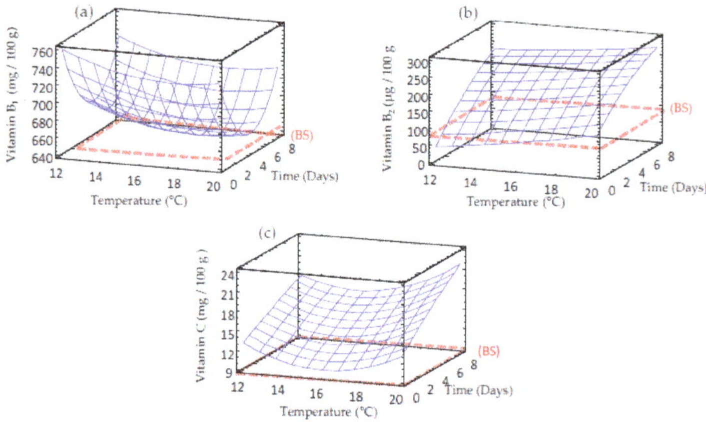


**Figure 4.** Response surface plots for vitamin content in germinated barley flour as a function of two independent variables (time and temperature). (**a**) Vitamin $B_1$; (**b**) Vitamin $B_2$ (**c**) Vitamin C. BS: barley seed flour (non-germinated).

### 3.4. Effect of Germination on β-Glucan Content of Sprouted Barley Flours

The β-glucan content was determined in raw and germinated barley (Table 2 and Figure 5a). The results show that β-glucan levels in barley grains were close to 5 g/100 g d.m. and decreased to a short extent during germination for short times (reductions of 14%–22% with respect to the initial content after 84-86 h of germination), but longer germination times cause greater losses. The impact of germination time on β-glucan levels was stronger at lower temperatures. It has been reported that germination causes an increase of β-glucanases activity, which are the primary enzymes responsible for breaking down cereal endosperm cell walls during germination, thus reducing β-glucan levels [28,29]. Our results agree with those reported by Rimsten et al. [30], who showed slight losses in β-glucan levels in barley grains germinated for 96 h after steeping at 48 °C. On the contrary, larger reductions of β-glucan content (50% after 48–120 h of germination) were found in germinated barley [31]. These studies have shown that longer germination times caused higher β-glucan degradation, results in line with the ones found in the present study.

### 3.5. Effect of Germination on GABA Content of Sprouted Barley Flours

Non-sprouted barley showed a GABA content of around 54 mg/100 g d.m. that significantly increased during germination (Table 2, Figure 5b), reaching values between 81 and 186 mg/100 g d.m. in sprouted barley, depending on germination conditions. The increase of GABA content in barley sprouts can be attributed to the partial hydrolysis of storage proteins to oligopeptides and free amino acids that are used for seedlings growth during germination, followed by the activation of glutamate decarboxylase enzyme that converts glutamic acid to GABA [10,32]. GABA can also be synthesized from polyamines by diamine oxidase (DAO), whose activity has been reported to increase during grain germination [33]. The enhancement of GABA content during cereal germination has been also observed by other authors in rice, barley and wheat [9,10,34].

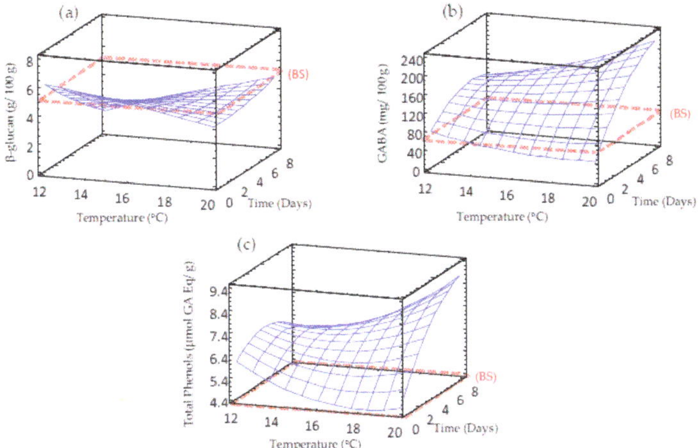


**Figure 5.** Response surface plots for bioactive compounds content in germinated barley flour as a function of two independent variables (time and temperature). (**a**) β-glucan; (**b**) GABA; (**c**) TPC. BS: barley seed flour (non-germinated).

### 3.6. Effect of Germination on Total Phenolic Content of Sprouted Barley Flours

One of the main advantages of germination in controlled conditions is the possibility of enhancing the antioxidant properties of grains by increasing phenolics contents.

The total phenolic contents (TPCs) increased significantly in barley after germination compared to the control, regardless of the conditions used during the process (Table 2 and Figure 5c). TPC levels ranged from 4.4 to 9.5 µmol GA Eq/g d.m. in sprouted barley. Germination time showed a significant linear impact on TPC content, especially at higher temperatures. The TPC levels in raw barley were in agreement with the values reported by Zhao et al. [35], who reported contents between 3.07 and 4.48 µmol GA Eq/g d.m. in different barley varieties. The increased levels of these bioactive compounds during barley sprouting is associated to structural changes of barley matrix that cause the release of phenolic compounds bound to cellular structures, and also to glycosylation reactions during germination, enhancing the extraction of free phenolic acids [36,37].

### 3.7. Effect of Germination on Free and Bound Phenolic Compounds Profile of Sprouted Barley Flours

Table 3 shows the content of free and bound phenolic compounds in barley seed and sprouts obtained at different conditions. Free phenolic fraction was mainly constituted of flavanols such as catechin and its dimeric form procyanidin B, while hydroxycinnamic acids, namely ferulic acid, 4-coumaric acid and caffeic acid, were the most abundant bound phenols. Non-germinated barley seeds contained a high concentration of procyanidin B (12.44 ± 0.80 µg/mg of extract) and catechin (0.99 ± 0.16 µg per mg of extract) in the soluble fraction (free phenolics), while the insoluble fraction (bound phenolics) was abundant in ferulic acid (45.06 ± 3.09 µg/mg of extract), 4-coumaric acid, (2.34 ± 0.40 µg/mg of extract), procyanidin B (1.06 ± 0.14 µg per mg of extract) and 4-hydroxybenzoic acid (0.45 ± 0.01 µg/mg of extract). The results found in this study are in agreement with those previously reported in barley [38,39].

**Table 3.** Content of free and bound phenolic compounds (μg/mg extract) in germinated barley flour as a function of two independent variables (time and temperature).

| Fraction | Germination Experiments | Catechin | Protocatechuic Acid | Diferulic Acid | 4-hydroxybenzoic Acid | Phloridzin | Ferulic Acid | Caffeic Acid | 4-coumaric Acid | Procyanidin B |
|---|---|---|---|---|---|---|---|---|---|---|
| **Free** | **BS** | $0.99 \pm 0.16^{cB}$ | $0.01 \pm 0.02^{aA}$ | $2.22 \pm 0.31^{dC}$ | ND $^{aA}$ | $0.01 \pm 0.00^{aA}$ | ND $^{aA}$ | ND $^{aA}$ | ND $^{aA}$ | $12.44 \pm 0.80^{cD}$ |
| | **1** | $0.20 \pm 0.04^{bB}$ | $0.02 \pm 0.00^{bA}$ | $0.76 \pm 0.32^{cC}$ | ND $^{aA}$ | $0.02 \pm 0.01^{aA}$ | ND $^{aA}$ | ND $^{aA}$ | ND $^{aA}$ | $1.94 \pm 0.56^{abB}$ |
| | **2** | $0.02 \pm 0.00^{aA}$ | $0.02 \pm 0.00^{bA}$ | $0.36 \pm 0.19^{bB}$ | ND $^{aA}$ | $0.02 \pm 0.01^{aA}$ | ND $^{aA}$ | ND $^{aA}$ | ND $^{aA}$ | $3.26 \pm 1.41^{bC}$ |
| | **3** | $0.53 \pm 0.40^{bcB}$ | $0.01 \pm 0.02^{aA}$ | $1.19 \pm 0.02^{eB}$ | ND $^{aA}$ | $0.03 \pm 0.01^{aA}$ | ND $^{aA}$ | ND $^{aA}$ | ND $^{aA}$ | $1.40 \pm 0.57^{abB}$ |
| | **4** | $0.12 \pm 0.12^{aA}$ | $0.03 \pm 0.00^{bA}$ | $1.06 \pm 0.07^{deB}$ | ND $^{aA}$ | $0.01 \pm 0.00^{aA}$ | ND $^{aA}$ | ND $^{aA}$ | ND $^{aA}$ | $1.06 \pm 0.01^{aB}$ |
| | **5** | $0.18 \pm 0.02^{bB}$ | $0.02 \pm 0.00^{bA}$ | $1.61 \pm 0.84^{defC}$ | ND $^{aA}$ | $0.02 \pm 0.01^{aA}$ | ND $^{aA}$ | ND $^{aA}$ | ND $^{aA}$ | $1.68 \pm 1.11^{abC}$ |
| | **6** | $0.67 \pm 0.71^{bcAB}$ | $0.03 \pm 0.00^{bA}$ | $1.90 \pm 0.82^{efB}$ | ND $^{aA}$ | $0.01 \pm 0.00^{aA}$ | ND $^{aA}$ | ND $^{aA}$ | ND $^{aA}$ | $2.79 \pm 2.23^{aB}$ |
| | **7** | $0.08 \pm 0.08^{aA}$ | $0.03 \pm 0.00^{bA}$ | $0.65 \pm 0.00^{cB}$ | ND $^{aA}$ | $0.01 \pm 0.00^{aA}$ | ND $^{aA}$ | ND $^{aA}$ | ND $^{aA}$ | $0.49 \pm 0.70^{aAB}$ |
| | **8** | $0.87 \pm 0.75^{bcB}$ | $0.02 \pm 0.00^{bA}$ | $1.03 \pm 0.10^{cdB}$ | ND $^{aA}$ | $0.01 \pm 0.00^{aA}$ | ND $^{aA}$ | ND $^{aA}$ | ND $^{aA}$ | $2.68 \pm 2.05^{abB}$ |
| | **9** | $0.57 \pm 0.36^{bcB}$ | $0.03 \pm 0.00^{bA}$ | $0.99 \pm 0.02^{dB}$ | ND $^{aA}$ | $0.02 \pm 0.00^{aA}$ | ND $^{aA}$ | ND $^{aA}$ | ND $^{aA}$ | $6.72 \pm 2.87^{bC}$ |
| **Bound** | **BS** | ND $^{aA}$ | $0.07 \pm 0.05^{bcB}$ | $0.26 \pm 0.05^{bD}$ | $0.45 \pm 0.01^{dE}$ | $0.02 \pm 0.01^{aA}$ | $45.06 \pm 3.09^{dH}$ | $0.16 \pm 0.01^{bC}$ | $2.34 \pm 0.40^{cdG}$ | $1.07 \pm 0.14^{aF}$ |
| | **1** | ND $^{aA}$ | $0.10 \pm 0.00^{cB}$ | $0.09 \pm 0.00^{aB}$ | $0.31 \pm 0.02^{cC}$ | $0.01 \pm 0.00^{aA}$ | $36.45 \pm 2.44^{cF}$ | $0.70 \pm 0.09^{dC}$ | $2.46 \pm 0.30^{cdE}$ | $1.22 \pm 0.18^{aD}$ |
| | **2** | $0.02 \pm 0.03^{aA}$ | $0.03 \pm 0.01^{abA}$ | $0.09 \pm 0.00^{aB}$ | $0.21 \pm 0.04^{bC}$ | $0.01 \pm 0.00^{aA}$ | $50.27 \pm 9.18^{deF}$ | $0.13 \pm 0.01^{bB}$ | $2.32 \pm 0.43^{cdE}$ | $1.07 \pm 0.33^{aD}$ |
| | **3** | ND $^{aA}$ | $0.08 \pm 0.06^{bcB}$ | $0.19 \pm 0.15^{abBC}$ | $0.22 \pm 0.02^{bC}$ | $0.02 \pm 0.01^{aA}$ | $31.15 \pm 1.89^{cE}$ | $0.33 \pm 0.19^{bcC}$ | $1.56 \pm 0.27^{bcD}$ | $1.23 \pm 0.29^{aD}$ |
| | **4** | ND $^{aA}$ | $0.05 \pm 0.00^{bB}$ | $0.19 \pm 0.02^{bC}$ | $0.27 \pm 0.09^{bcC}$ | $0.02 \pm 0.01^{aA}$ | $31.25 \pm 5.16^{cF}$ | $0.22 \pm 0.01^{cC}$ | $1.92 \pm 0.19^{cd.E}$ | $0.90 \pm 0.03^{aD}$ |
| | **5** | ND $^{aA}$ | $0.06 \pm 0.01^{bB}$ | $1.18 \pm 0.1^{eD}$ | $0.15 \pm 0.08^{bB}$ | $0.01 \pm 0.01^{aA}$ | $23.82 \pm 1.02^{bF}$ | $0.67 \pm 0.01^{dC}$ | $1.40 \pm 0.03^{bE}$ | $0.66 \pm 0.94^{aBCD}$ |
| | **6** | ND $^{aA}$ | $0.08 \pm 0.02^{bc.B}$ | $1.52 \pm 0.70^{defCD}$ | $0.22 \pm 0.18^{bcB}$ | $0.01 \pm 0.00^{aA}$ | $28.40 \pm 17.43^{bcdE}$ | $0.68 \pm 0.41^{cdC}$ | $2.01 \pm 1.06^{bcdD}$ | $0.89 \pm 0.06^{aC}$ |
| | **7** | ND $^{aA}$ | $0.04 \pm 0.03^{abA}$ | $1.03 \pm 0.11^{dC}$ | $0.12 \pm 0.07^{bB}$ | $0.01 \pm 0.01^{aA}$ | $31.27 \pm 9.38^{cdE}$ | $0.26 \pm 0.10^{cB}$ | $1.79 \pm 0.57^{bcD}$ | $1.08 \pm 0.14^{aC}$ |
| | **8** | ND $^{aA}$ | $0.15 \pm 0.07^{cB}$ | $1.21 \pm 1.12^{bcdeCD}$ | $0.41 \pm 0.01^{dC}$ | $0.02 \pm 0.01^{aA}$ | $72.95 \pm 12.73^{eF}$ | $1.21 \pm 0.21^{eD}$ | $2.60 \pm 0.51^{cdE}$ | $0.99 \pm 0.07^{aD}$ |
| | **9** | ND $^{aA}$ | $0.13 \pm 0.03^{c}$ | $2.88 \pm 0.42^{fE}$ | $0.32 \pm 0.09^{cC}$ | $0.01 \pm 0.01^{aA}$ | $43.89 \pm 6.59^{dF}$ | $0.98 \pm 0.05^{eD}$ | $3.28 \pm 0.85^{dE}$ | $1.01 \pm 0.07^{aD}$ |

The results are the mean ± std ($n = 3$). Different superscript letters (small) within a column indicate significant differences ($p \leq 0.05$) among germination treatments. Different superscript letters (capital) within a row indicate significant differences among individual phenolic compounds. Codes 1–9 corresponded to the germination experiments performed according to the RSM design (Table 1). BS: Barley seed flour (non-germinated). ND: not detected.

The content of ferulic acid in the bound fraction increased during sprouting as the germination time augmented at higher temperatures (Table 2 and Figure 6a), which might be attributed to *de novo* synthesis in the embryonic axis of sprouted grain [40]. Our results align with those previously observed in wheat by other authors [41,42], who found an increase of ferulic acid levels after germination for 24–96 h. As expected, the content of procyanidin B decreased during germination (Figure 6b), followed by a rise in catechin content. The decrease of procyanidin B levels might be associated with its degradation to individual catechin monomers to protect the seed at this very vulnerable growth stage.

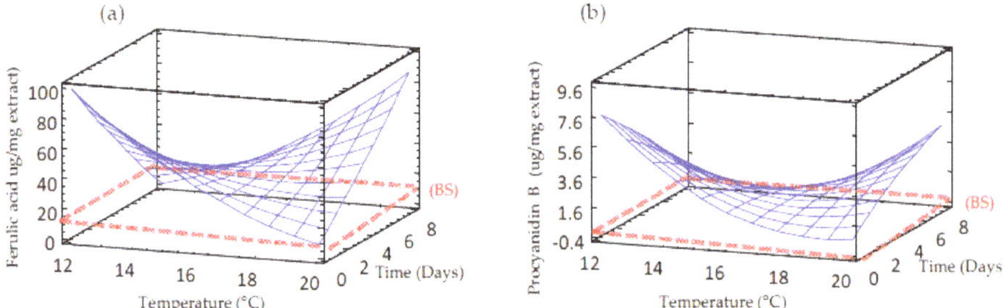


**Figure 6.** Response surface plots for predominant free and bound phenolic compounds in germinated barley flour as a function of two independent variables (time and temperature). (**a**) Ferulic acid; (**b**) Procyanidin B. BS: non-germinated barley flour.

### 3.8. Effect of Germination on Antioxidant Activity of Sprouted Barley Flours

Total antioxidant activity was measured as a comprehensive assessment through the evaluation of different antioxidant parameters (Table 2 and Figure 7). The germination process produced an enhancement of the antioxidant activity, which was mainly time-dependent, exhibiting the sprouts produced at longer germination times the highest antioxidant activities, regardless of the assay used for evaluation (FRAP, ORAC, DPPH, TEAC and Q-DPPH). The increase in antioxidant potential in sprouted barley could be attributed to the release of phenolic compounds from the cell wall components during germination process, compounds that can act as hydrogen or electron donors [43]. Since DPPH scavenging activity was also evaluated without a previous extraction step, this antioxidant activity in sprouted barley cannot be solely attributed to the increase of free phenolic compounds during germination. These findings suggest that other antioxidant compounds present in barley sprouts, such as vitamin C, also contribute to its high antioxidant potential.

### 3.9. Effect of Germination on Glycemic Index (GI) of Sprouted Barley Flours

The GI indicates the rate at which 50 g of carbohydrate in a particular food is absorbed into the blood stream as blood sugar and ranges from 1 to 100. Glucose is used as reference food and is rated 100. Foods can be classified by their GI, according to Gordillo-Bastidas et al. [44], as follows: (i) High GI foods (>70), (ii) Medium GI foods (between 56–69) and iii) Low GI foods (≤55).

GI was evaluated in all raw and sprouted barley samples (Table 2, Figure 8). As germination proceeded, GI increased significantly in barley as consequence of the release of reducing sugars. In fact, studies performed in barley and brown rice found an enhancement of reducing sugars levels during germination due to hydrolysis of starch by α-amylase and β-amylase enzymes [31,45].

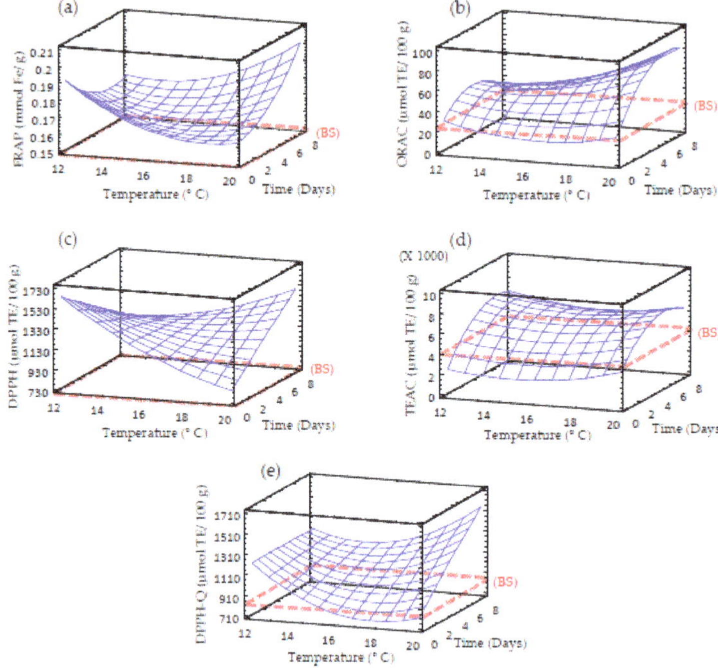


**Figure 7.** Response surface plots for antioxidant activity in germinated barley flour as a function of two independent variables (time and temperature). (**a**) FRAP; (**b**) ORAC; (**c**). DPPH; (**d**) TEAC; (**e**) DPPH-Q. BS: barley seed flour (non-germinated).

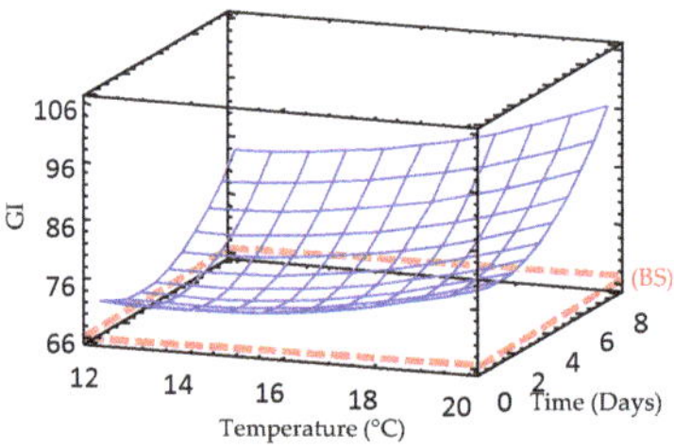


**Figure 8.** Response surface plots for glycemic index in germinated barley flour as a function of two independent variables (time and temperature). BS: barley seed flour (non-germinated).

### *3.10. Effect of Germination on the Colour and Sensory Properties of Sprouted Barley Flours*

Two phenomena were observed when the colour of germinated barley flours was compared with that of raw barley flour (Table 2, Figures 9 and 10). Germination decreased the brightness (L*) of barley flour. However, redness (a*) and yellowness (b*) increased as the germination proceed (Table 2). These results can be attributed to the starch and protein hydrolysis during germination and, subsequently, the generation of Maillard reaction compounds during drying treatment. Moreover, during the soaking period, there is a migration of bran pigments into the water and into the endosperm. Moreover, heating treatments might affect the brown barley colour. These results are in agreement with the results observed by the sensory panel (Table 2 and Figure 11) who observed an increase in colour change associated with the loss of brightness and increased flavor intensity in germinated barley. Texture was also modified, as an increasing coarseness of the powder with germination temperature was observed.

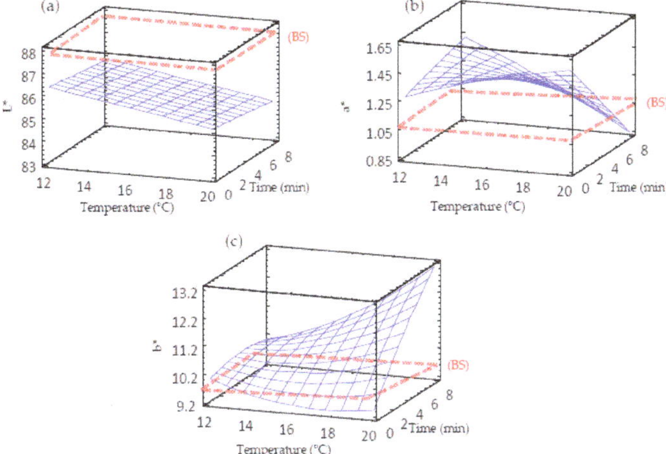


**Figure 9.** Response surface plots for color in germinated barley flour as a function of two independent variables (time and temperature). (**a**) L* value; (**b**) a* value; (**c**) b* value. BS: barley seed flour (non-germinated).

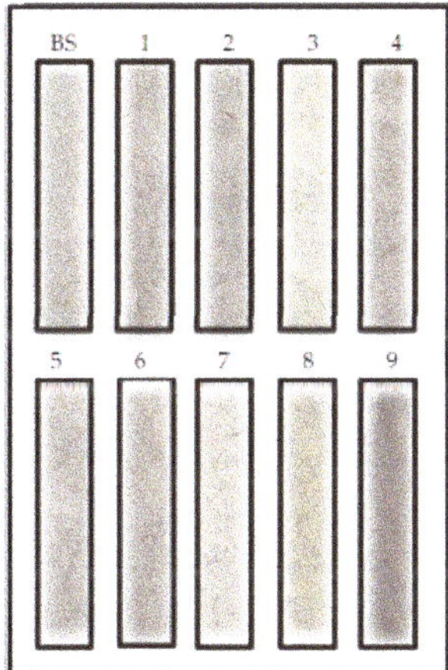


**Figure 10.** Colour of flours from germinated barley flour obtained at different conditions of time and temperature. BS: barley seed flour (non-germinated). Codes 1 to 9 corresponded to the germination experiments performed according to RSM design (Table 1).

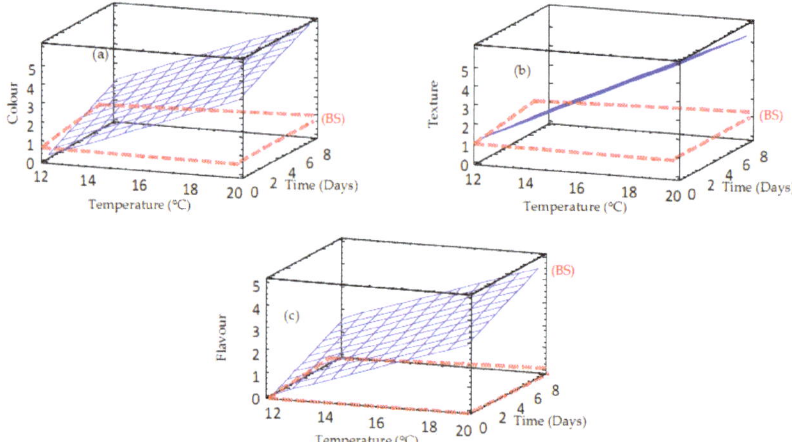


**Figure 11.** Response surface plots for sensory properties of germinated barley flour as a function of two independent variables (time and temperature). (**a**) Colour; (**b**) Texture; (**c**) Flavour. BS: non-germinated barley flour.

### 3.11. Multivariate Optimization of Germination Conditions to Obtain High-Quality Germinated Barley Flours

The desirability function was used to select the optimal germination conditions to obtain the highest predicted values for nutritional and bioactive compounds as well as for antioxidant activity and the lowest values for saturated fatty acid content and glycemic index. Based on this criterion, the most suitable germination conditions were temperature and time of 16 °C and 3.53 days, respectively, conditions that provide an overall desirability value of 0.44 (Figure 12).

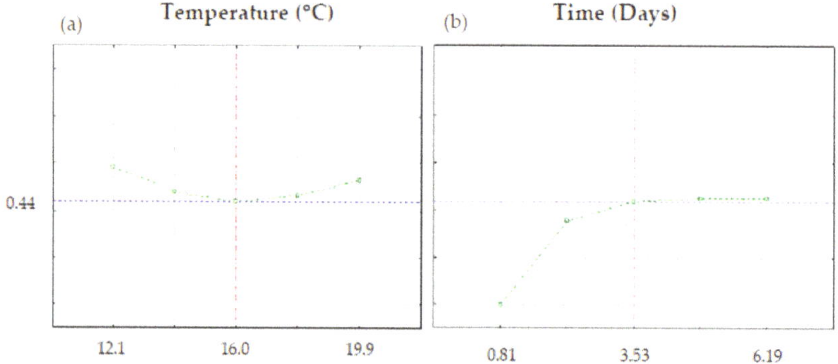


**Figure 12.** Optimal germination conditions determined by desirability function. (**a**) Optimal temperature; (**b**) Optimal time.

## 4. Conclusions

This study demonstrates that germination is a promising process for developing novel nutritive and functional flours from barley with improved quality features. An increase in germination time caused an enhancement of the nutritional and bioactive potential of sprouted barley flour, but also an increase of carbohydrates hydrolysis and, consequently, a higher glycemic index. Since the aim of this study was to produce improved barley flours from the nutritional and functional point of view, the use of mild temperatures (16 °C) and short times (3.53 days) could be promising to produce nutritive barley flours with high antioxidant properties and a low glycemic index, which may be beneficial for consumers´ health.

**Author Contributions:** Conceptualization, D.R., E.P., C.M.-V., J.F. and A.B.M.-D.; Methodology, D.R., E.P., M.d.C.G., R.I.B., D.K.R. and A.B.M.-D.; Validation, D.R., E.P., C.M.-V., J.F. and A.B.M.-D.; Formal analysis, D.R., E.P., M.d.C.G. and A.B.M.-D.; Investigation, D.R., E.P., M.d.C.G., C.M.-V., R.I.B., D.K.R., J.F. and A.B.M.-D.; Resources, D.R. and A.B.M.-D.; Data curation, D.R., E.P., C.M.-V., J.F. and A.B.M.-D.; Writing—original draft preparation, D.R., E.P. and A.B.M.-D.; Writing—review and editing, D.R., E.P., C.M.-V., R.I.B., D.K.R., J.F. and A.B.M.-D.; Visualization, D.R., E.P., C.M.-V., R.I.B., D.K.R., J.F. and A.B.M.-D.; Supervision D.R., E.P., C.M.-V., J.F. and A.B.M.-D.; Project administration, D.R., E.P. and A.B.M.-D.; Funding acquisition, D.R., E.P., C.M.-V., J.F., and A.B.M.-D. All authors have read and agreed to the published version of the manuscript.

**Funding:** This research was funded by Agrotechnological Institute (ITACyL) (Castilla y León. Spain) and the European Union through FEADER, grant number Biodough 2018-807, European Union through FEDER, grant number Invfood 2018-400, (MINECO/FEDER), grant number AGL2015-67598-R and CSIC, intramural project grant number 201870I097.

**Acknowledgments:** The authors acknowledge Dr. Francisco Ciudad Bautista for providing barley variety obtained in ITACyL, IRTA, EEDF-CSIC, ITAP and INIA (1FD97-0792 and RTA2006-00020-C04).

**Conflicts of Interest:** The authors declare no conflict of interest.

## References

1. Madakemohekar, A.; Prasad, L.C.; Pal, J.P.; Prasad, R. Estimation of combining ability and heterosis for yield contributing traits in exotic and indigenous crosses of barley (Hordeum vulgare L.). *Res. Crop.* **2018**, *19*, 264–270. [CrossRef]
2. Martínez, M.; Motilva, M.J.; de las Hazas, M.C.L.; Romero, M.P.; Vaculova, K.; Ludwig, I.A. Phytochemical composition and β-glucan content of barley genotypes from two different geographic origins for human health food production. *Food Chem.* **2018**, *245*, 61–70. [CrossRef]
3. Commission (EU) Regulation No 1048/2012 of 8 November 2012: On the Authorisation of a Health Claim Made on Foods and Referring to the Reduction of Disease Risk. Available online: https://eurex.europa.eu/LexUriServ/LexUriServ.do?uri=OJ:L:2012:310:0038:0040:EN:PDF (accessed on 17 February 2020).
4. Ötman, E.M.; Frid, A.H.; Groop, L.C.; Bjorck, I.M.E. A dietary exchange of common bread for tailored bread of low glycaemic index and rich in dietary fibre improved insulin economy in young women with impaired glucose tolerance. *Eur. J. Clin. Nut.* **2005**, *60*, 334–341. [CrossRef]
5. Šimić, G.; Horvat, D.; Lalić, A.; Komlenić, D.K.; Abičić, I.; Zdunić, Z. Distribution of β-glucan, phenolic acids, and proteins as functional phytonutrients of hull-less barley grain. *Foods* **2019**, *8*, 680. [CrossRef]
6. Madakemohekar, A.; Prasad, L.C.; Prasad, R. Generation mean analysis in barley (*Hordeum vulgare* L.) under drought stress condition. *Plant Arch.* **2018**, *18*, 917–922.
7. Holtekjølen, A.K.; Olsen, H.R.; Færgestad, E.M.; Uhlen, A.K.; Knutsen, S.H. Variations in water absorption capacity and baking performance of barley varieties with different polysaccharide content and composition. *LWT Food Sci. Technol.* **2008**, *41*, 2085–2091. [CrossRef]
8. Vicentini, A.; Liberatore, L.; Mastrocola, D. Functional foods: Trends and development of the global market. *Ital. J. Food. Sci.* **2016**, *28*, 338–351.
9. Cáceres, P.J.; Martínez-Villaluenga, C.; Amigo, L.; Frias, J. Maximising the phytochemical content and antioxidant activity of Ecuadorian brown rice sprouts through optimal germination conditions. *Food Chem.* **2014**, *152*, 407–414. [CrossRef]
10. Cáceres, P.J.; Peñas, E.; Martinez-Villaluenga, C.; Amigo, L.; Frias, J. Enhancement of biologically active compounds in germinated brown rice and the effect of sun-drying. *J. Cereal Sci.* **2017**, *73*, 1–9. [CrossRef]
11. Frias, J.; Peñas, E.; Vidal-Valverde, C. Changes of vitamin content of powder enteral formulas as a consequence of storage. *Food Chem.* **2009**, *115*, 1411–1416. [CrossRef]
12. Martin-Diana, A.B.; Izquierdo, N.; Albertos, I.; Sanchez, M.S.; Herrero, A.; Sanz, M.A.; Rico, D. Valorization of Carob's germ and seed peel as natural antioxidant ingredients in gluten-free crackers. *J. Food Process. Preserv.* **2017**, *41*, e12770. [CrossRef]
13. Mattila, P.; Pihlava, J.M.; Hellström, J. Contents of phenolic acids, alkyl- and alkenylresorcinols, and avenanthramides in commercial grain products. *J. Agric. Food Chem.* **2005**, *53*, 8290–8295. [CrossRef]

14. Birsan, R.; Wilde, P.; Waldron, K.; Rai, D. Recovery of polyphenols from brewer's spent grains. *Antioxidants* **2019**, *8*, 380. [CrossRef]
15. Brand-Williams, W.; Cuvelier, M.E.; Berset, C. Use of a free radical method to evaluate antioxidant activity. *LWT Food Sci. Technol.* **1995**, *28*, 25–30. [CrossRef]
16. Re, R.; Pellegrini, N.; Proteggente, A.; Pannala, A.; Yang, M.; Rice-Evans, C. Antioxidant activity applying an improved ABTS radical cation decolorization assay. *Free Radic. Biol. Med.* **1999**, *26*, 1231–1237. [CrossRef]
17. Ou, B.; Hampsch-Woodill, M.; Prior, R.L. Development and validation of an improved oxygen radical absorbance capacity assay using fluorescein as the fluorescent probe. *J. Agric. Food Chem.* **2001**, *49*, 4619–4626. [CrossRef]
18. Vijayalakshmi, M.; Ruckmani, K. Ferric reducing anti-oxidant power assay in plant extract. *Bangladesh J. Pharmacol.* **2016**, *11*, 570–572. [CrossRef]
19. Gularte, M.; Rosell, C. Physicochemical properties and enzymatic hydrolysis of different starches in the presence of hydrocolloids. *Carbohyd. Polym.* **2011**, *85*, 237–244. [CrossRef]
20. Granfeldt, Y. Foods Factors Affecting Metabolic Responses to Cereal Products. Ph.D. Thesis, University of Lund, Lund, Sweden, 1994.
21. Onimawo, I.A.; Asugo, S. Effects of germination on the nutrient content and functional properties of pigeon pea flour. *J. Food Sci. Technol.* **2004**, *41*, 170–174.
22. Gómez-Favela, M.A.; Gutiérrez-Dorado, R.; Cuevas-Rodríguez, E.O.; Canizalez-Román, V.A.; del Rosario, C.L.-S.; Milán-Carrillo, J. Improvement of chia seeds with antioxidant activity, GABA, essential amino acids, and dietary fiber by controlled germination bioprocess. *Plant Food Hum. Nutr.* **2017**, *72*, 345–352. [CrossRef]
23. Chavarín-Martínez, C.D.; Gutiérrez-Dorado, R.; Perales-Sánchez, J.X.K.; Cuevas-Rodríguez, E.O.; Milán-Carrillo, J.; Reyes-Moreno, C. Germination in optimal conditions as effective strategy to improve nutritional and nutraceutical value of underutilized Mexican blue maize seeds. *Plant Food Hum. Nutr.* **2019**, *74*, 192–199. [CrossRef] [PubMed]
24. Krist, S.; Stuebiger, G.; Unterweger, H.; Bandion, F.; Buchbauer, G. Analysis of volatile compounds and triglycerides of seed oils extracted from different poppy varieties (*Papaver somniferum* L.). *J. Agric. Food Chem.* **2005**, *53*, 8310–8316. [CrossRef] [PubMed]
25. Copeland, L.; McDonald, M.B. Seed germination. In *Principle of Seed Science and Technology*, 4th ed.; Copeland, L., Ed.; Springer: Cham, Switzerland, 2001; pp. 72–124.
26. Bibi, N.; Aurang, Z.; Amal, B.K.; Mohammad, S.K. Effect of germination time and type of illumination on proximate composition of chickpea seed (*Cicer arietinum* L.). *Am. J. Food Technol.* **2008**, *3*, 24–32. [CrossRef]
27. Arora, S.; Jood, S.; Khetarpaul, N. Effect of germination and probiotic fermentation on nutrient composition of barley based food mixtures. *Food Chem.* **2010**, *119*, 779–784. [CrossRef]
28. Marconi, O.; Tomasi, I.; Dionisio, L.; Perretti, G.; Fantozzi, P. Effects of malting on molecular weight distribution and content of water-extractable β-glucans in barley. *Food Res. Int.* **2014**, *64*, 677–682. [CrossRef]
29. Betts, N.S.; Berkowitz, O.; Liu, R.; Collins, H.M.; Skadhauge, B.; Dockter, C.; Burton, R.A.; Whelan, J.; Fincher, G.B. Isolation of tissues and preservation of RNA from intact, germinated barley grain. *Plant J.* **2017**, *91*, 754–765. [CrossRef]
30. Rimsten, L.; Haraldsson, A.-K.; Andersson, R.; Alminger, M.; Sandberg, A.-S.; Åman, P. Effects of malting on glucanase and phytase activity in barley grain. *J. Sci. Food Agric.* **2002**, *82*, 904–912. [CrossRef]
31. Farzaneh, V.; Ghodsvali, A.; Bakhshabadi, H.; Zare, Z.; Carvalho, I.S. The impact of germination time on the some selected parameters through malting process. *Int. J. Biol. Macromol.* **2017**, *94*, 663–668. [CrossRef]
32. Chu, C.; Yan, N.; Du, Y.; Liu, X.; Chu, M.; Shi, J.; Zhang, H.; Liu, Y.; Zhang, Z. iTRAQ-based proteomic analysis reveals the accumulation of bioactive compounds in Chinese wild rice (*Zizania latifolia*) during germination. *Food Chem.* **2019**, *289*, 635–644. [CrossRef]
33. Yang, R.; Chen, H.; Gu, Z. Factors influencing diamine oxidase activity and γ-aminobutyric acid content of fava bean (*Vicia faba* L.) during germination. *J. Agric. Food Chem.* **2011**, *59*, 11616–11620. [CrossRef]
34. Chung, H.J.; Jang, S.H.; Cho, H.Y.; Lim, S.T. Effects of steeping and anaerobic treatment on GABA (γ-aminobutyric acid) content in germinated waxy hull-less barley. *LWT Food Sci. Technol.* **2009**, *42*, 1712–1716. [CrossRef]
35. Zhao, H.; Fan, W.; Dong, J.; Lu, J.; Chen, J.; Shan, L.; Lin, Y.; Kong, W. Evaluation of antioxidant activities and total phenolic contents of typical malting barley varieties. *Food Chem.* **2008**, *107*, 296–304. [CrossRef]

36. Leitao, C.; Marchioni, E.; Bergaentzlé, M.; Zhao, M.; Didierjean, L.; Miesch, L.; Holder, E.; Miesch, M.; Ennahar, S. Fate of polyphenols and antioxidant activity of barley throughout malting and brewing. *J. Cereal Sci.* **2012**, *55*, 318–322. [CrossRef]
37. Carvalho, D.O.; Gonçalves, L.M.; Guido, L.F. Overall antioxidant properties of malt and how they are influenced by the individual constituents of barley and the malting process. *Compr. Rev. Food Sci. Food Saf.* **2016**, *15*, 927–943. [CrossRef]
38. Carvalho, D.O.; Curto, A.F.; Guido, L.F. Determination of phenolic content in different barley varieties and corresponding malts by liquid chromatography-diode array detection-electrospray ionization tandem mass spectrometry. *Antioxidants* **2015**, *4*, 563–576. [CrossRef] [PubMed]
39. Gangopadhyay, N.; Rai, D.K.; Brunton, N.P.; Gallagher, E.; Hossain, M.B. Antioxidant-guided isolation and mass spectrometric identification of the major polyphenols in barley (hordeum vulgare) grain. *Food Chem.* **2016**, *210*, 212–220. [CrossRef]
40. Zilic, S.; Basic, Z.; Sucakovic, V.H.T.; Maksimovic, V.; Jancovic, M.; Filipovic, M. Can the sprouting process applied to wheat improve the contents of vitamins and phenolic compounds and antioxidant capacity of the flour? *Int. J. Food Sci. Technol.* **2014**, *49*, 1040–1047. [CrossRef]
41. Yang, F.; Basu, T.K.; Ooraikul, B. Studies on germination conditions and antioxidant contents of wheat grain. *Int. J. Food Sci. Nutr.* **2001**, *52*, 319–330. [CrossRef]
42. Kim, M.J.; Kwak, H.C.S.; Kim, S.S. Effects of germination on protein, $\gamma$-aminobutyric acid, phenolic acids, and antioxidant capacity in wheat. *Molecules* **2018**, *23*, 2244. [CrossRef]
43. Fernandez-Orozco, R.; Frias, J.; Zielinski, H.; Piskula, M.K.; Kozlowska, H.; Vidal-Valverde, C. Kinetic study of the antioxidant compounds and antioxidant capacity during germination of *Vigna radiata* cv. emmerald, *Glycine max* cv. jutro and *Glycine max* cv. merit. *Food Chem.* **2008**, *111*, 622–630. [CrossRef]
44. Gordillo-Bastidas, E.; Díaz-Rizzolo, D.A.; Roura, E.; Massanés, T.; Gomis, R. Quinoa (*Chenopodium quinoa* wild), from nutritional value to potential health benefits: An integrative review. *J. Nutr. Food Sci.* **2016**, *6*, 1–10.
45. Charoenthaikij, P.; Jangchud, K.; Jangchud, A.; Prinyawiwatkul, W.; No, H. Composite wheat–germinated brown rice flours: Selected physicochemical properties and bread application. *Int. J. Food Sci. Technol.* **2012**, *47*, 75–82. [CrossRef]

*Article*

# Exploiting Milling By-Products in Bread-Making: The Case of Sprouted Wheat

**Gaetano Cardone, Paolo D'Incecco, Maria Cristina Casiraghi and Alessandra Marti ***

Department of Food, Environmental and Nutritional Sciences (DeFENS), Università degli Studi di Milano, 20133 Milan, Italy; gaetano.cardone@unimi.it (G.C.); paolo.dincecco@unimi.it (P.D.); maria.casiraghi@unimi.it (M.C.C.)

* Correspondence: alessandra.marti@unimi.it

Received: 11 February 2020; Accepted: 25 February 2020; Published: 1 March 2020

**Abstract:** This research investigated the effect of sprouting on wheat bran. Bran from un-sprouted (BUW) and sprouted (BSW) wheat were characterized in terms of chemical composition, enzymatic activities, and hydration properties. In addition, the rheological properties (using GlutoPeak, Farinograph, Extensograph, and Rheofermentometer tests) and bread-making performance (color, texture, volume of bread) of wheat doughs enriched in bran at 20% replacement level were assessed. Sprouting process caused a significant decrease in phytic acid (~20%), insoluble dietary fiber (~11%), and water holding capacity (~8%), whereas simple sugars (~133%) and enzymatic activities significantly increased after processing. As regards the gluten aggregation kinetics, the BSW-blend profile was more similar to wheat than BUW-blend, indicating changes in the fiber and gluten interactions. BSW led to a worsening of the mixing and leavening properties, instead, no significant changes in extensibility were observed. Finally, BSW improved bread volume (~10%) and crumb softness (~52%). Exploiting bran from sprouted wheat might be useful to produce bread rich in fiber with enhanced characteristics.

**Keywords:** bran; cell walls; sprouting; dough rheology; bread-making; microstructure

## 1. Introduction

Fiber-enrichment of food products has become increasingly important as a means to increase their nutritional properties. In this context, bran from cereals-with a total dietary fiber content of 30%–50%—is one of the most important source of dietary fiber used in the bread-making industry [1]. However, the inclusion of high levels of fiber in cereal-based products remains a technological challenge, due to the need to maintain acceptable dough rheological properties as well as sensory attributes. Indeed, adding high levels of bran to dough leads to an increase in water absorption, a decrease in both mixing stability and leavening tolerance [2,3]. The most evident effects on the final baked product are the decrease in loaf volume, the increase in crumb firmness, the appearance of dark crumb, and, in some cases, the modification of taste with the appearance of bitterness [4].

The detrimental effect of bran addition on bread-making cannot be solely attributed to the dilution of gluten proteins and to the physical disruption of gluten network, but the physical, chemical, and biochemical properties of bran should be also considered [5]. Besides specific physical properties—i.e., the strong tendency of bran to absorb water that might result in competition for water between bran and other key flour components like starch and proteins—bran seems to have a certain chemical reactivity (i.e., between ferulic acid and proteins) which might determine its functionality [5].

Several pre-treatments have been proposed to counter these negative effects, such as: (i) particle size reduction, which significantly influences the rheological properties of dough in terms of mixing time, stability and dough resistance to extension [6], (ii) application of high-pressure [7], and (iii) enzymatic treatment [8], which alters the physical and structural properties of dough and its interaction with water and (iv) fermentation, which improves the bioactivity and baking properties of dough enriched with wheat bran [9]. Considering all these treatments, the best results were obtained when exogenous enzymes were used as such [6] or produced by microorganisms [9].

In this context, sprouting (or germination) can be proposed as a bio-technological process able to promote the accumulation of enzymatic activities. A recent study reported that wholegrain flour from sprouted wheat could be used to produce bread with improved characteristics (i.e., specific volume and crumb softness) compared to conventional wholegrain flours [10,11]. Enhancements in bread attributes were also found by using refined flour from sprouted wheat [12,13]. Specifically, Marti et al. [12] proposed the use of flour from sprouted wheat as alternative to the conventional enzymatic improvers in bread-making. Indeed, low amount of sprouted wheat flour (<2%) enhanced the bread-making attitude of stiff flour, with the advantage of producing a clean label product.

Considering the potential use of flour from sprouted wheat, it remains to elucidate the physical, chemical, and structural aspects of the wheat milling by-products (i.e., bran) and how these characteristics might affect the baking performance of bran-enriched wheat bread. Indeed, although in the literature there is a large amount of study available on the enrichment of baked products in bran, no information is available on the use of bran from sprouted wheat in bread-making. Starting from this point, the purpose of this research was to investigate the features of bran obtained from sprouted wheat and how this ingredient affects both dough rheological properties and bread characteristics.

## 2. Materials and Methods

### 2.1. Materials

Wheat kernels (*Triticum aestivum* L.) were kindly supplied by Molino Quaglia (Molino Quaglia S.p.A., Vighizzolo d'Este, Italy). A part of kernels (10 tons) was sprouted at industrial scale for 48 h, and then dried at 60 °C for 12 h, as previously reported by Marti et al. [13]. Both samples, un-sprouted and sprouted kernels, were milled using a laboratory mill (Labormill, BONA, Monza, Italy) to collect bran (bran yield: 20%).

Wheat brans (bran from un-sprouted and sprouted wheat—BUW and BSW—respectively) were toasted (Self-Cooking Center®—Rational AG, Landsberg am Lech, Germany) at 200 °C for 120 s, in order to inactivate parts of the enzymes. Bran fractions with particle size >500 μm were further ground using the Cyclotec 1093 (FOSS, Höganäs, Sweden) to decrease their size (<500 μm). BUW and BSW were used in replacement of the 20% of a commercial refined wheat flour (CTRL; $W = 280 \times 10^{-4}$ J; P/L = 1.16) provided by Molino Quaglia (Molino Quaglia S.p.A., Vighizzolo d'Este, Italy). In this way, two whole-wheat flours were obtained which differences were associated solely to bran type (BUW or BSW).

### 2.2. Analytical Methods

#### 2.2.1. Microstructural Evaluation

Kernels from either un-sprouted or sprouted wheat were prepared for confocal laser scanning microscopy (CLSM) and light microscopy (LM) to specifically evaluate changes due to sprouting process itself. Specimens were fixed and dehydrated according to Faltermaier et al. [14] then embedded in a methacrylate resin (Technovit 7100, Wertheim, Germany). After resin polymerization, 10 μm-thick sections were obtained using a rotary microtome (Leitz 1512). Sections were treated, with 2,4-dinitrophenylhydrazine (20 min) followed by washing in tap water (30 min), then in 0.5% periodic acid (20 min) and again in tap water (30 min) [14]. This procedure is a modification of the periodic acid–Schiff's reaction that allows the dye acid fuchsin to be specific for protein without binding to other polysaccharides like starch.

The staining for protein and cell walls was performed by using 0.1% (*w*/*v*) water solution of acid fuchsin (Sigma-Aldrich, St Louis, MO, USA) for 1 min and 10% (*w*/*v*) Calcofluor white for 1 min (fluorescent brightener 28, Sigma-Aldrich, Milan, Italy), respectively. Sections were inspected using an inverted CLSM (Nikon A1+, Minato, Japan). Acid fuchsin was excited at 560 nm and the emission filter was set at 630–670 nm while Calcofluor white was excited at 409 nm and the emission filter was set at 430–480 nm. Starch was stained by using Lugol's solution (5 g $I_2$ and 10 g KI in 100 mL MilliQ water), and sections were examined after 5 min staining with an Olympus BX light microscope (Tokyo, Japan) equipped with QImaging Retiga camera (Surrey, Canada).

BUW and BSW were prepared for CLSM as mentioned above and observed exploiting the auto-fluorescence of the samples. Specimens for SEM observations were prepared according to Cardone et al. [10] observed using a Zeiss LEO 1430 SEM at 3 kV.

#### 2.2.2. Chemical Composition

Moisture content of bran was evaluated at 130 °C until the sample weight did not change by 1 mg for 60 s in a moisture analyzer (Radwag—Wagi Elektroniczne, Chorzòw, Poland). Total starch content was evaluated by standard method (AACC 76-13.01) [15]. Sugars were assessed by HPLC by Anion Exchange Chromatography with Pulsed Amperometric Detection (HPAEC-PAD) [16]. Total (TDF), soluble (SDF), and insoluble (IDF) dietary fiber contents were quantified by an enzymatic-gravimetric procedure as reported by standard method (AOAC 991.43) [17].

Total and soluble arabynoxylans were determined as reported by Manini et al. [18]. The phytic acid content was determined by HPLC with spectrophotometric detection as previously reported by Oberleas and Harland [19].

#### 2.2.3. Enzymatic Activities

α-amylase activity of bran was evaluated by following standard method (AACC 22-02.01) [15]. The analysis of xylanase activity was performed by using the Azo-wheat arabinoxylan kit (K-AZOWAX 09/04; Megazyme International Ireland Ltd., Wicklow, Ireland) with some modifications (i.e., 1 h of incubation instead of 20 min).

#### 2.2.4. Hydration Properties

The water holding capacity (WHC) of bran was determined as reported by Lebesi and Tzia [20] by suspending 0.5 g of each bran samples with 45 mL distilled water. Instead, water binding capacity (WBC) was evaluated according to Zanoletti et al. [21].

#### 2.2.5. Rheological Properties

The effects of bran on flour functionality and bread-making performances were determined on two blends prepared by substituting the 20% of commercial flour with BUW and BSW at the same level, respectively. Control flour without bran was also examined.

##### Gluten Aggregation Properties

The aggregation properties of gluten were investigated by means of the GlutoPeak (Brabender GmbH and Co. KG, Duisburg, Germany) test, according to Marti et al. [12]. The major indices considered were: (i) Maximum Torque (MT, expressed in Brabender Equivalents, BE), indicating the peak following the aggregation of the gluten proteins; (ii) Peak Maximum Time (PMT, expressed in s), indicating the time to obtain the Maximum Torque; (iii) Total Energy (expressed in GlutoPeak Equivalent, GPE) indicating the area under the curve from the beginning of the analysis up to 15 s after the Maximum Torque.

##### Mixing Properties

Mixing properties were evaluated following to standard method (ICC 115/1) [22] in the 50 g kneading bowl of the Farinograph-E® (Brabender GmbH & Co. KG, Duisburg, Germany).

Extensibility Properties

The extensograph test was carried out on a 20 g dough by means of the micro-Extensograph® (Brabender GmbH & Co. KG, Duisburg, Germany), at three resting times (45, 90, and 135 min). Dough samples were prepared by following to standard method (AACC 54-10.01) [15], in the 50 g kneading bowl of the Farinograph-E® (Brabender GmbH & Co. KG, Duisburg, Germany).

Leavening Properties

The Rheofermentometer® (Chopin, Tripette, and Renaud, Villeneuve La Garenne Cedex, France) was used to analyze dough development and carbon dioxide ($CO_2$) production and retention using the method described by Marti et al. [13].

### 2.2.6. Baking Test

A straight-dough method was applied for the production of bread according to Cardone et al. [10]. The amount of water added to the sample and the mixing time of the dough varied for each formulation according to previously determined farinographic indices. The dough samples were split into sub-samples of 80 g each, formed into cylindrical shapes, put into square shape bread molds (height: 4 cm; length: 9 cm; depth: 6 cm) and left to leaven for 60 min in a thermostatic chamber at 30 °C (70% relative humidity). To prevent the formation of crust too quickly, the leavened dough was baked in an oven (Self Cooking Center®, Rational International AG, Landsberg am Lech, Germany) in two stages. Firstly, the samples were baked at 120 °C with vapor injection (90% relative humidity) for 4 min and then the oven temperature was increased to 230 °C for 11 min. The resulting loaves, two hours after baking, were analyzed or packaged in orientated polypropylene film for three days.

### 2.2.7. Bread Properties

Color

The evaluation of the browning (100-L*) and saturation of the color intensity (redness and yellowness, a* and b*, respectively) of bread crust and crumb were assessed with a reflectance color meter (CR 210, Minolta Co., Osaka, Japan).

Specific Volume

The bread specific volume was performed through the ratio between volume and mass of bread and expressed in mL/g. The apparent volume was evaluated by the sesame displacement approach.

Crumb Hardness

Crumb hardness was assessed by means of a dynamometer (Z005, Zwick Roell, Ulm, Germany), equipped with a 100 N load cell as previously described by Marti et al. [23]. Bread samples were evaluated two hours after baking (day zero), and after one and three days of storage. Crumb hardness was determined as the maximum compression force at a deformation of 30%. Three central slices (15 mm thick) of one loaf from each bread-making trial were analyzed. The values of hardness resulting from each slice of bread were analyzed by applying the Avrami Equation (1):

$$\theta = \frac{(A_\infty - A_t)}{(A_\infty - A_0)} = e^{-ktn} \tag{1}$$

where $\theta$ indicates the fraction of the total change in hardness still to occur, $A_0$, $A_t$, and $A_\infty$ are experimental values of hardness at times zero, $t$ and infinity respectively, $k$ is a rate constant and $n$ ($n$ = 1) is the Avrami exponent. All the parameters were obtained from the modelling process.

### 2.3. Statistics

Chemical composition, enzymatic activities, hydration properties, and gluten aggregation properties were determined in triplicate, whereas mixing and leavening properties were analyzed in duplicate. As regards the extensograph test, the determinations for each sample were made in duplicate and from each dough two subsamples were analyzed. For the bread production, two baking tests were carried out, and three loaves were prepared from each baking test. Color measurements were replicated five times. Bread specific volume, crumb porosity, and hardness were carried out on six loaves or slices.

To determine differences between means (for BUW and BWS) a paired t-test was applied. Analysis of variance (one-way ANOVA) was performed by utilizing the Fisher's Least Significant Difference (LSD) test. Data were elaborated by Statgraphic XV v. 15.1.02 (StatPoint Inc., Warrenton, VA, USA).

## 3. Results

### 3.1. Microstructure Evaluation of Kernels and Brans before and after Sprouting

Microstructural modifications due to sprouting process were evaluated on the whole kernel to avoid interferences of mechanical breaks possibly brought by milling. Sprouting caused the hydrolysis of the subaleurone endosperm cell walls, that were no longer visible in sprouted kernels (Figure 1A,B), and the hydrolysis of some starch granules near the aleurone cells (head arrows in Figure 1B), consistently with the hydrolytic enzyme synthesis in the aleurone layer. Fluorescence of Calcofluor white, specific for cell walls especially β-1,4-glucans, considerably decreased in sprouted kernels (Figure 1C,D). At the same time, the decrease in fluorescence of acid fuchsin suggested a protein degradation in subaleurone region (Figure 1C,D).

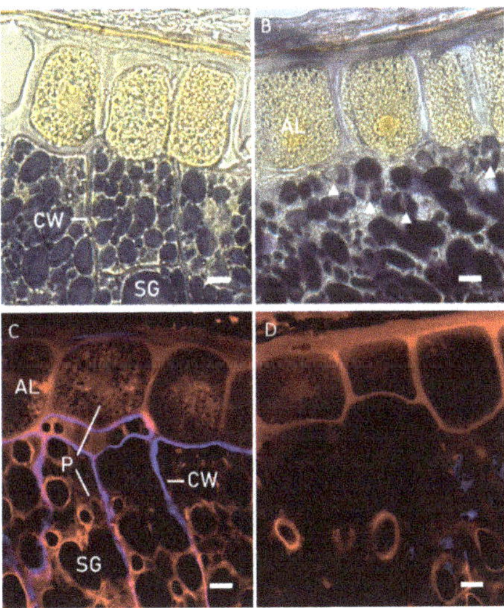


**Figure 1.** Light microscopy (**A**,**B**) and confocal laser scanning microscopy (**C**,**D**) of kernels before (**A**–**C**) and after (**B**–**D**) sprouting process. AL: aleuron layer, CW: cell wall, SG: starch granule, P: protein. White head arrows in panel B show hydrolyzed starch granules. Bars are 10 μm in length.

CLSM of brans (Figure 2A,B) clearly show that BUW is almost composed of pericarp and aleurone layers, whereas BSW is richer in starch granules since large portions of starchy endosperm remained adherent to the aleurone layer. This difference may be a consequence of different break paths after sprouting and it explains the lower flour yield of BSW (51% vs. 54%; data not shown). SEM analysis confirmed the relevant structural changes induced by sprouting and showed the hydrolysis of starch granules mainly to occur in the BSW as erosion pits (red arrow in Figure 2E).

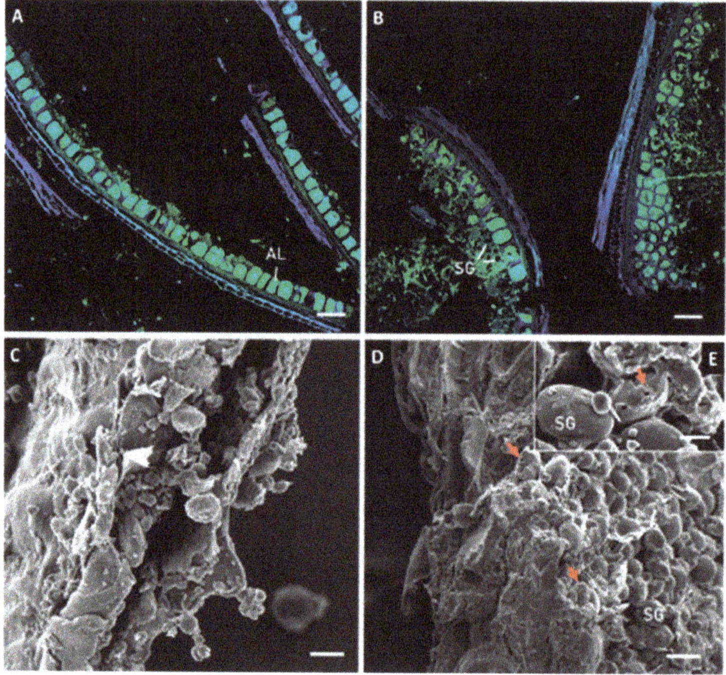


**Figure 2.** Confocal laser scanning microscopy (**A**,**B**) and scanning electron microscopy (**C–E**) of bran before (**A–C**) and after (**B**,**D**,**E**) sprouting process. AL: aleuron layer, SG: starch granule. Red arrows in panel D and E show the effect of amylolytic enzymes as "erosion pits" on the granule surface. Bars are 40 μm in length in panels A,B, 10 μm in length in panels C,D and 5 μm in panel E.

### 3.2. *Effect of Sprouting on Bran Features*

Sprouting process increased the amount of total sugars in bran, and specifically glucose, fructose and maltose (Table 1). The higher starch content in BSW might be the consequence of the different compactness and, therefore, milling behavior of BUW and BSW, as shown in Figure 2. A significant decrease in IDF was found after sprouting (−11% in BSW respect to BUW). Instead, the SDF content of BUW and BSW was not affected by the process. Moreover, sprouting did not induce a significant degradation and solubilization of arabinoxylans. On the contrary, in the case of prolonged sprouting (i.e., four days), a significant decrease in the total content of arabinoxylans and an increase in water-extractable ones was reported in barley [24]. In contrast, the sprouting degraded antinutritive factors, such as phytic acid, confirming previous studies carried out on wholemeal flours [25].

**Table 1.** Effect of sprouting on wheat bran characteristics.

| | Bran from Unsprouted Wheat (BUW) | Bran from Sprouted Wheat (BSW) |
|---|---|---|
| Total starch | 18.3 ± 0.5 | 26.9 ± 2.0 ** |
| Arabinoxylans | | |
| Total | 14.4 ± 0.5 | 12.0 ± 0.7 |
| Soluble | 0.22 ± 0.01 | 0.29 ± 0.08 |
| Phytic acid | 13.98 ± 0.46 | 11.29 ± 0.02 * |
| Total Sugar | 3.0 ± 0.9 | 7.0 ± 2.2 * |
| Glucose | 0.14 ± 0.01 | 0.53 ± 0.02 * |
| Fructose | 0.05 ± 0.01 | 0.20 ± 0.00 * |
| Sucrose | 2.02 ± 0.20 | 4.99 ± 0.00 * |
| Raffinose | 0.79 ± 0.02 | n.d. |
| Maltose | n.d. | 1.29 ± 0.03 |
| Total Dietary Fiber (TDF) | 45.4 ± 0.6 | 40.2 ± 0.9 * |
| Soluble (SDF) | 2.2 ± 0.4 | 2.0 ± 0.3 |
| Insoluble (IDF) | 43.1 ± 0.2 | 38.2 ± 0.6 * |
| α-amylase activity | 0.094 ± 0.003 | 143 ± 16 * |
| Xylanase activity | 0.16 ± 0.04 | 0.35 ± 0.03 * |
| Water Holding Capacity | 4.9 ± 0.1 | 4.5 ± 0.1 * |
| Water Binding Capacity | 3.9 ± 0.2 | 3.7 ± 0.2 |

Values associated with asterisks in the same row are significantly different (*t*-test, * $p \leq 0.05$; ** $p \leq 0.001$); n.d.: not detectable. Compositional and hydration property data are expressed as g/100g sample (d.b.). α-amylase activity and xylanase activity are expressed as Ceralpha Units (CU)/g flour and as activity/g flour, respectively. TDF, total dietary fiber; SDF, soluble dietary fiber; IDF, insoluble dietary fiber.

As expected, sprouting promoted an accumulation of both α-amylases and xylanases in the bran fraction (Table 1). Specifically, the activity of α-amylase in BSW increased by about 1500-fold compared to BUW, whereas the xylanase activity in BSW was 1.2-fold higher than BUW.

BUW and BSW hydration properties were assessed in terms of their water holding (WHC) and water binding (WBC) capacity. The WHC significantly decreased after sprouting process, by about 8% (Table 1). Instead, there was no significant differences between BUW and BSW in terms of WBC.

### 3.3. Gluten Aggregation and Mixing Properties

During the GlutoPeak test, the speed of the rotating paddle allows the formation of gluten, and a rapid increase in the torque curve occurs. Additional mixing breaks down the gluten network and the torque curve declines [26]. The gluten aggregation kinetics of CTRL was typical of a strong flour with good bread-making performance that is usually characterized by long aggregation time (i.e., PMT), high torque (i.e., MT) and high energy values (Figure 3 and Table 2).

Replacing 20% of refined wheat flour with both types of bran, the PMT and MT indices significantly decreased and increased, respectively, resulting a decrease in the energy value, suggesting gluten weakening [27]. Worsening in gluten aggregation properties where more evident when BUW was added to CTRL sample, instead of BSW.

The CTRL flour used for making the blends was characterized by a long dough development time and high stability (Figure 4; Table 2), in agreement with the GlutoPeak test (Figure 3 and Table 2). When a 20% of flour was replaced by BUW or BSW, a significant increase in water absorption was observed (Table 2), with BUW-blend absorbing more water than BSW-blend, in agreement with the hydration properties of the related bran samples (Table 1). Regarding the development time, the presence of BUW decreased the time needed to achieve maximum consistency (−48%). This effect was even more evident when BSW was added (−58%). Dough stability, which in wheat dough indicates dough strength, decreased significantly by about 47% and 72% in the case of BUW- and BSW-blends, respectively. In contrast to the GlutoPeak test, the farinograph test showed the highest weakening of the dough when BSW was added.

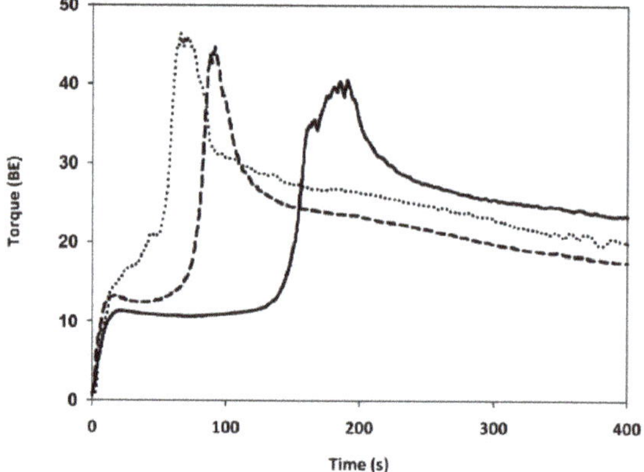


**Figure 3.** GlutoPeak profiles of refined wheat flour alone (solid line) and in presence of bran from un-sprouted wheat (dotted line) or bran from sprouted wheat (dash line).

**Table 2.** Gluten aggregation and mixing properties of refined wheat flour alone, with bran from un-sprouted wheat, or bran from sprouted wheat.

| | | | Refined Wheat Flour (CTRL) | CTRL + BUW | CTRL + BSW |
|---|---|---|---|---|---|
| GLUTEN AGGREGATION PROPERTIES (GlutoPeak Test) | Peak maximum time | (s) | 192 ± 3 [c] | 65 ± 1 [a] | 93 ± 4 [b] |
| | Maximum torque | (BE) | 40.3 ± 0.6 [a] | 45.7 ± 0.6 [c] | 43.7 ± 2 [b] |
| | Total Energy | (GPE) | 3567 ± 112 [c] | 1914 ± 25 [a] | 2059 ± 35 [b] |
| MIXING PROPERTIES (Farinograph Test) | Water absorption | (%) | 58.1 ± 0.2 [a] | 65.6 ± 0.5 [c] | 61.8 ± 0.5 [b] |
| | Development time | (min) | 10.8 ± 0.3 [c] | 5.6 ± 0.3 [b] | 4.5 ± 0.5 [a] |
| | Stability | (min) | 23 ± 2 [c] | 12 ± 2 [b] | 6.4 ± 0.07 [a] |
| | Degree of softening | (FU) | 15 ± 1 [a] | 44 ± 5 [b] | 103.5 ± 0.7 [c] |

At different letters (a, b, c) in the same row, correspond significant differences (one-way ANOVA, LSD test, $p \leq 0.05$). BUW: bran from unsprouted wheat; BSW: bran from sprouted wheat. Peak maximum time: indicates the time to obtain the Maximum Torque; Maximum torque: indicates the peak following the aggregation of the gluten proteins; Energy: indicates the area under the curve from the beginning of the analysis up to 15 s after the Maximum Torque. BE: Brabender Equivalent; GPE: GlutoPeak Equivalent. Water absorption: amount of water to obtain a dough with optimal consistency (500 ± 20 FU); Dough development time: time required for the dough development; Stability: point between arrival time (when the top of the curve reaches the 500 FU line) and departure time (when the top of the curve leaves the 500 FU line); Degree of softening: difference between the 500 FU line and the dough consistency at 12 min after the dough development. FU: Farinograph Units.

### 3.4. Dough Extensibility

Dough extensibility significantly decreased when bran was added, with no significant differences according to the type of bran (Figure 5a–c and Table 3). In addition, all samples showed a decrease in dough extensibility as the resting time increased, in particular when the resting time was extended from 45 min to 90 min. However, the extensibility for the CTRL flour remained practically unmodified between 90 and 135 min.

Bran-enrichment did not cause any significant modification regarding resistance at 45 and 90 min of resting time, while after 135 min of resting time, bran caused an increase in dough resistance, suggesting dough stiffness [28]. This phenomenon is more evident when BSW was included in the dough.

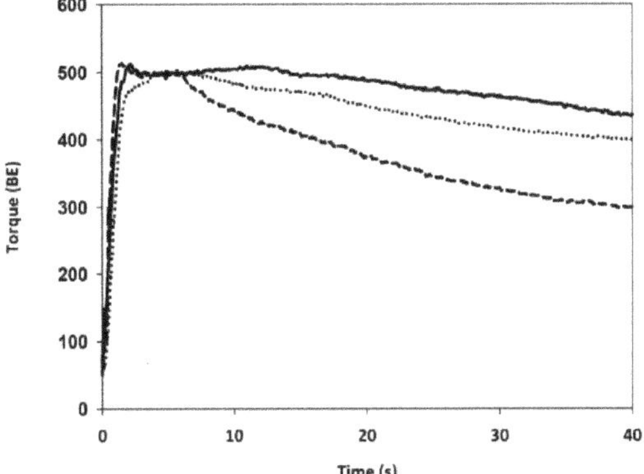


**Figure 4.** Farinograph profiles of refined wheat flour alone (solid line) and in presence of bran from un-sprouted wheat (dotted line) or bran from sprouted wheat (dash line).

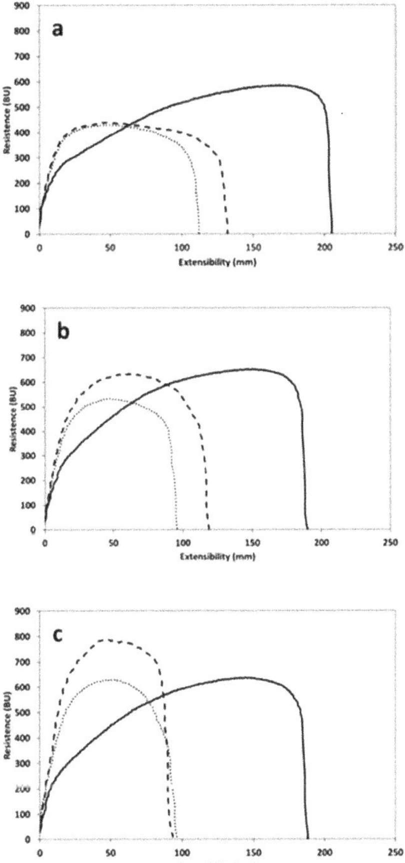


**Figure 5.** Micro-Extesograph profiles of refined wheat flour alone (solid line) and in presence of bran from un-sprouted wheat (dotted line) or bran from sprouted wheat (dash line), after 45 min (**a**), 90 min (**b**), and 135 min (**c**) of resting time.

**Table 3.** Extensibility properties of refined wheat flour alone, with bran from un-sprouted wheat, or bran from sprouted wheat.

| | Resting Time | CTRL | CTRL + BUW | CTRL + BSW |
|---|---|---|---|---|
| Extensibility (E) (mm) | 45 min | 209 ± 11 [b] | 119 ± 6 [a] | 132 ± 4 [a] |
| Resistance to extension (R) (BU) | | 353 ± 46 [a] | 423 ± 29 [a] | 431 ± 1 [a] |
| R/E ratio | | 1.7 ± 0.3 [a] | 4.0 ± 0.4 [b] | 3.3 ± 0.1 [b] |
| Energy ($cm^2$) | | 155 ± 9 [b] | 76 ± 2 [a] | 88 ± 2 [a] |
| Extensibility (E) (mm) | 90 min | 182 ± 7 [b] | 104 ± 6 [a] | 109 ± 3 [a] |
| Resistance to extension (R) (BU) | | 405 ± 63 [a] | 512 ± 26 [a,b] | 583 ± 41 [b] |
| R/E ratio | | 2.2 ± 0.3 [a] | 4.9 ± 0.5 [b] | 5.4 ± 0.2 [b] |
| Energy ($cm^2$) | | 147 ± 31 [b] | 77 ± 4[a] | 94 ± 11 [a,b] |
| Extensibility (E) (mm) | 135 min | 185 ± 2 [b] | 97 ± 2 [a] | 97 ± 1 [a] |
| Resistance to extension (R) (BU) | | 430 ± 45[a] | 609 ± 32 [b] | 721 ± 7 [c] |
| R/E ratio | | 2.3 ± 0.3 [a] | 6.0 ± 0.3 [b] | 7.5 ± 0.1 [c] |
| Energy ($cm^2$) | | 163 ± 4 [b] | 85 ± 5 [a] | 101 ± 3 [a] |

At different letters (a, b, c) in the same row, correspond significant differences (one-way ANOVA, LSD test, $p \leq 0.05$). CTRL: refined wheat flour; BUW: bran from unsprouted wheat; BSW: bran from sprouted wheat. E: dough extensibility; R: resistance to constant deformation after 50 mm stretching; R/E: ratio of resistance to extension and extensibility.

The ratio number, which indicates the ratio between the resistance to extension and extensibility, increased with the addition of bran, confirming the role of this milling fraction in inducing dough stiffness. This phenomenon is well shown by the high ratio number values at 135 min. The highest increase in ratio number was recorded when BSW was added to the CTRL flour. Finally, the energy required for deformation was decreased by the addition of bran, particularly when BUW was used.

### 3.5. Leavening Properties

Maximum dough height during leavening was not significantly altered by the addition of bran (Figure 6). On the other hand, the time of maximum dough development decreased, from 180 min in the control to 161 min and 131 min by adding BUW and BSW, respectively. The time of dough development and the loss in dough volume (weakening coefficient), was high for all bran-enriched samples, especially when BSW was added. The decrease in dough stability agrees with the farinograph index. The drop off after three hours of leavening ranged from 6% and 17.5% for BUW and BSW samples.

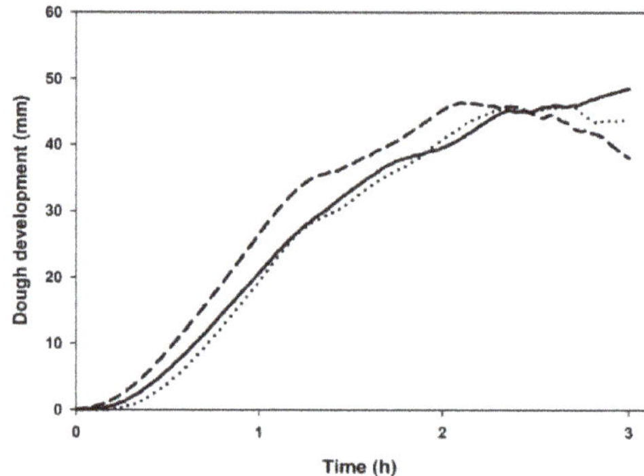


**Figure 6.** Dough development profiles of commercial wheat flour (solid line); with bran from un-sprouted wheat (dotted line) or bran from sprouted wheat (dash line).

Regarding gas production and retention (Table 4), dough with bran produced more gas than CTRL and this is due to a higher content of simple sugars (Table 1), which are consumed faster by yeast compared to CTRL. This effect was more noticeable when BSW was used. As regards the gas retention coefficient (Table 4), the BUW addition did not induce a significant negative effect after three hours of leavening. In contrast, in presence of BSW the gas retention capacity decreased from 94% (CTRL) to 86%, after the same time (i.e., three hours).

**Table 4.** Leavening properties of refined wheat flour alone, with bran from un-sprouted wheat, or bran from sprouted wheat.

| | CTRL | CTRL + BUW | CTRL + BSW |
|---|---|---|---|
| Maximum dough height (mm) | 51 ± 3 [b] | 47 ± 1 [a] | 47 ± 1 [a] |
| Time of maximum dough development (min) | 180 ± 0 [c] | 161 ± 2 [b] | 131 ± 6 [a] |
| Dough height at 180 min (mm) | 51 ± 3 [b] | 44 ± 1 [a] | 39 ± 1 [a] |
| Weakening coefficient at 180 min (%) | n.d. | 6 ± 1 [b] | 17.5 ± 0.4 [a] |
| Total $CO_2$ (mL) | 1129 ± 52 [a] | 1301 ± 42 [a,b] | 1476 ± 84 [b] |
| Retained $CO_2$ (mL) | 1064 ± 45 [a] | 1197 ± 29 [b] | 1271 ± 46 [b] |
| Released $CO_2$ (mL) | 65 ± 7 [a] | 104 ± 12 [a] | 205 ± 38 [b] |
| $CO_2$ retention coefficient (%) | 94 ± 1 [b] | 92 ± 0.1 [b] | 86 ± 2 [a] |
| Porosity time (min) | n.d. | 103 ± 4 [b] | 76 ± 4 [a] |

At different letters (a, b, c) in the same row, correspond significant differences (one-way ANOVA, LSD test, $p \leq 0.05$). n.d.: not detectable. CTRL: refined wheat flour; BUW: bran from unsprouted wheat; BSW: bran from sprouted wheat. Maximum dough height: maximum height reached during the leavening; final dough height: height reached by the dough at the end of the analysis; maximum height: maximum height relative to gas production; porosity time: time when the dough porosity appears; total $CO_2$: total amount of $CO_2$ produced; $CO_2$ retained: amount of $CO_2$ retained in the dough; $CO_2$ released: amount of $CO_2$ released from the dough; $CO_2$ retention coefficient: ratio between $CO_2$ retained and total $CO_2$.

### 3.6. Bread Properties

Bran-enriched bread resulted in a darker (increase in 100-L*) and redder (higher a*) crumb and crust than CTRL bread (Table 5). As expected, this phenomenon was more intense in BSW-enriched bread. When BSW was added instead of BUW, also the color crumb (Table 5) showed a significant increase in browning (100-L*) and in redness (a*) levels.

**Table 5.** Properties of bread from refined wheat flour alone, with bran from un-sprouted wheat, or bran from sprouted wheat.

| | | CTRL | CTRL + BUW | CTRL + BSW |
|---|---|---|---|---|
| Bread | Specific volume (mL/g) | 2.7 ± 0.6 [b] | 2.2 ± 0.1 [a] | 2.4 ± 0.1 [a,b] |
| Crust | Browning (100-L*) | 53 ± 4 [a] | 60 ± 4 [b] | 60 ± 1 [b] |
| | Redness (a*) | 9 ± 3 [a] | 12 ± 1 [b] | 12 ± 1 [b] |
| | Yellowness (b*) | 25 ± 2 [b] | 20 ± 5 [a] | 19 ± 1 [a] |
| Crumb | Browning (100-L*) | 48 ± 3 [a] | 52 ± 2 [b] | 57 ± 2 [c] |
| | Redness (a*) | −1.3 ± 0.2 [a] | 3.80 ± 0.3 [c] | 3.2 ± 0.4 [b] |
| | Yellowness (b*) | 18 ± 1 [b] | 16 ± 1 [a] | 16 ± 1 [a] |
| Crumb firming kinetics | $A_0$ (N) | 6.48 | 6.62 | 3.12 |
| | $A_\infty$ (N) | 66.01 | 34.55 | 31.35 |
| | $A_\infty$-$A_0$ (N) | 59.53 | 27.93 | 28.23 |
| | $k$ ($h^{-n}$) | 0.186 | 0.480 | 0.252 |

At different letters (a, b, c) in the same row, correspond significant differences (one-way ANOVA, LSD test, $p \leq 0.05$). CTRL: refined wheat flour; BUW: bran from unsprouted wheat; BSW: bran from sprouted wheat. $A_0$: hardness at times zero; $A_\infty$: hardness at infinity; $k$: rate constant; n: Avrami exponent.

As expected, the addition of bran had a slight negative effect on specific volume, however no significant differences were found between CTRL and BSW.

The addition of BSW to wheat resulted in a relevant decrease in crumb hardness (3.1 N) compared to the other samples (6.0 N). During storage (up to 3 days), only BSW-samples exhibited higher softness than the CTRL (Figure 7). In addition, the softness of BSW-enriched bread after 3 days was similar in softness to the CTRL and BUW-enriched breads after one day.

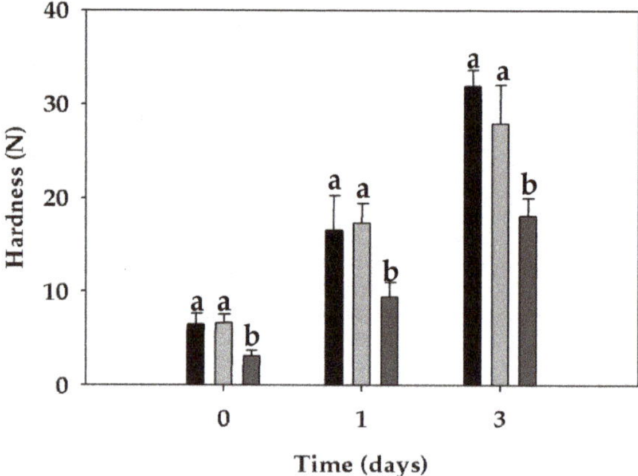


**Figure 7.** Crumb firmness properties of loaves obtained from refined wheat flour (black), with bran from un-sprouted wheat (light grey), or bran from sprouted wheat (dark grey) during storage. Values associated with different letters (a, b) in the same day are significantly different (one-way ANOVA, LSD test, $p \leq 0.05$).

To evaluate the kinetic attitude of starch during the retrogradation phenomena, the Avrami equation was applied, fixing the value of Avrami exponent ($n = 1$) as proposed by Michniewicz et al. [29]. The estimated Avrami coefficient $k$ for bran-enriched bread samples was higher than that for the CTRL sample. The high coefficient of starch retrogradation with BUW described a fast firming rate. The BSW led to a decrease in the firming rate (Table 5) compared to the BUW and CTRL. Regarding the firmness at infinite time ($A_\infty$) the highest value was found for the CTRL sample, instead the lowest $A_\infty$ was observed in the presence of BSW. The intensity of staling evaluated as the total firmness increment ($A_\infty$–$A_0$) decreased with the addition of bran, due to the high water binding capacity of fiber. For this parameter, no difference was observed between bran-enriched breads.

## 4. Discussion

Due to the recent interest in using refined flour from sprouted wheat in bread-making as flour improver or ingredient [12,13], it is worthy of interest to investigate the potential use of bran obtained from the milling of sprouted wheat. In this study, the effect of sprouting was assessed on both physical and chemical properties of wheat bran. Moreover, the effect of wheat replacement (at 20% level which is roughly the percentage of bran in wholegrain) with bran from sprouted wheat was studied on both dough rheology and bread-making performance.

Despite the well-known nutritional benefits of bran, the presence of antinutritional factors (i.e., phytic acid) inhibit the absorption of minerals and vitamins [30]. Several authors have been reported the positive effects of sprouting process on the degradation of phytates in wheat and other cereals [25]. As reported in Table 1, the sprouting conditions applied in the present study caused a significant decrease in phytic acid.

From the technological standpoint, it is well known that the incorporation of fiber into flour negatively affects the textural and sensory properties of cereal-based products. Indeed, numerous negative effects on dough and breads properties have been as attributed to bran, including increased dough stickiness, decreased mixing, fermentation tolerance, volume, and crumb softness [4]. For this reason, it is indispensable to modify the structural properties of fiber (i.e., by increasing the soluble fraction) to improve the quality of bran-enriched foods. Specifically, the soluble fraction of the dietary fiber contributes to the creation of the viscosity of the liquid phase of the system [31]. The sprouting effects on fiber content are reported to be varied and unclear, with increased, decreased and no change reported for several grains [32]. The sprouting process carried out under the conditions applied in this study determined a slight decrease in the TDF and IDF content, instead no changes was observed in terms of SDF (Table 1). A similar trend in dietary fiber change was also found by Koehler et al. [33] up to 4 days of sprouting at about 20 °C. The decrease in TDF is likely due to the breakdown of the water extractable dietary fiber components to smaller molecules which are not precipitable in ethanol, thus not counted as dietary fiber by the method used [8,34]. On the other hand, the decrease in IDF might be due to the xylanase activity developed during sprouting (Table 1). Indeed, such enzymes is the main responsible for the hydrolysis of xylans, the principal component of hemicellulose (about 30% of cell walls) [35]. The changes in dietary fiber component ratio might affect the functional properties of foods, which are mainly related to fiber-water interactions (i.e., water holding capacity (WHC) and water binding capacity (WBC)). The WHC indicates the amount of water that fibers can absorb in the absence of an external force, instead, WBC represents the amount of water that remains bound to fiber after the application of an external force [36]. Sprouting process affected fiber-water interactions as result of the changes induced by the xylanase activity developed during the process (Table 1). Specifically, the decrease in WHC in BSW could be due to the decrease in IDF [37] and perhaps to the lower (even if not statistically significant) content of water-unsoluble arabinoxylans, compared to BUW (Table 1). Indeed, water-unsoluble arabinoxylans are characterized by a stronger WHC compared to the water-soluble ones [38]. The significant decrease in IDF and the slightly increase in water-soluble arabinoxylans in BSW could also explain its lower ability to absorb water in dough system (see farinograph index; Table 2) compared to BUW. Indeed, these two components are responsible for the WHC of the fiber-enrich foods [39]. The xylan-degradation might profoundly modify the bran attitude in bread-making, likely due to modification in water distribution caused by losing their strong water-holding capacity [40]. In order to improve the bran enriched-bread acceptance, generally hemicelluloses (e.g., endoxylanase) are used in bread-making [39], since the soluble arabinoxylans released contribute to forming a hydrated network, together with gluten-forming proteins [41].

To understand the effects of sprouting on fiber-protein interactions, the bran-enriched flours were assessed by means of GlutoPeak test. Generally, flour with good bread-making quality is characterized by a much slower buildup in dough consistency, require more time to reach peak consistency and show high maximum torque [26]. Regardless the type of bran, the profile of bran-enriched flour samples suggested a weakening of gluten network, since the presence of fiber causes deleterious effects on dough structure due to the dilution of the gluten matrix [3]. Several authors postulated that the worsening of dough attributes cannot be related to gluten dilution only [4], but also to physical, chemical, and/or biochemical bran properties [5]. As reported in Table 2, BSW-blend profile showed longer peak maximum time than BUW-blend, suggesting that sprouting attenuated the negative impact on gluten aggregation properties. A possible reason is that the higher insoluble content of dietary fiber in the BUW-blend allowed the formation of a viscous film that hindered the interaction among flour particles [42], preventing the formation of a strong gluten network like that of BSW-blend. In addition, even the decrease in phytic acid content in the BSW-sample could play a role in determining a less negative impact on gluten aggregation. Indeed, phytic acid might react with gluten-forming proteins worsening their aggregation capacity [43,44]. Last but not least, the amount of ferulic acid—whose interaction with proteins negatively impact the gluten formation [5]—decreased upon sprouting [45].

The gluten weakening resulted from the addition of bran was also observed in dough system, evaluated as mixing properties (i.e., dough development time and stability, and degree of softening) (Table 2). In this system, the addition of BSW had a greater impact than BUW (Table 2). This result might be due to proteases expressed in the bran layers during sprouting process, as shown by CLSM (Figure 1D). The effect of proteases is evident during the farinograph test that provides for long mixing time (about 20 min). In accordance, Marti et al. [12] reported that after 48 h of sprouting, proteases in refined flour increased from 0.66 to 1.43 unit*g$^{-1}$.

Despite the worsening of mixing properties, the enzymatic activity developed during sprouting did not further worsen the extensibility properties of the dough (Table 3). Indeed, differences between BUW- and BSW-blends were observed only after 135 min in terms of resistance to extension and energy (Table 3). The high capacity of the BSW-blend to maintain the extensibility properties could be attributed to the new interactions established between the gluten network and the fiber able to partially contrast the negative effect of the hydrolytic activity developed during the sprouting process and activated during the test (Table 1). Selinheimo et al. [46] reported that proteases—specifically laccase which could play some role in the pre-harvest sprouting—increase the maximum resistance and, at the same time, decrease the extensibility of the dough. Moreover, it was reported that the use of cell-wall degrading enzyme (xylanases) increases the resistance to extension in wheat doughs [46,47]. The parameter of resistance to extension is an indicator of dough-handling properties and it is positively correlated to dough volume, since doughs characterized by a high resistance to deformation show better performance in bread-making [48]. In that respect, the more resistant gluten of BSW-blend led to a bread characterized by higher volume than BUW, although not statistically significant (Table 5). We expect that differences between the samples will be highlighted whenever the bread-making trials will be carried out on larger scale (i.e., >250 g loaf), instead of micro-scale, as in this study, or using different leavening processes.

The improvement of bread in the presence of BSW (Table 5) can also be linked to the xylanase activity developed during the sprouting process (Table 1). Indeed, in bread-making, xylanases are commonly used to improve dough handling properties [49], and loaf volume [49,50].

As regards crumb firming rate, it was evaluated by using the Avrami equation (1). In general, samples with high firmness rate constant (k) values are characterized by fast crumb firming kinetics, whereas, low $k$ values lead to a slow firming kinetics. Both firmness rate constant and initial crumb firmness showed a decrease when BSW was used (see crumb firmness kinetic values; Table 5) instead of BUW. This result might indicate that the xylanase activity developed during sprouting process improved the initial crumb texture and its firming rate of the resulting bread (Table 5). Indeed, the positive effects of xylanases on bread are related to the cleavage of the backbone of arabinoxylans, with the consequent release of water and decrease in water-insoluble pentosane [51]. In addition to xylanases, also $\alpha$-amylases have positive effects on dough development and crumb staling. The greater presence of fermentable sugars in the BSW (Table 1)—consumable by yeasts for their growth and $CO_2$ production—might have contributed positively to the decrease in leavening time (Table 4) and to the greater volume of the corresponding bread (Table 5) [52]. The presence of BSW enhanced the textural properties of loaves, since the crumb hardness—already after two hours of baking—was significantly lower compared to the other samples. This difference is not due to either the crumb moisture, which was not statistically different among the samples (data not shown), or to the crumb porosity, as the BSW-bread was characterized by the lowest value of this index. On the contrary, the crumb texture improvement is due to the $\alpha$-amylase activity, whose effects have been demonstrated in several studies [53–55]. Furthermore, the high $\alpha$-amylase activity of the BSW had positive effects on firming rate of the corresponding bread crumb (Table 5).

In conclusion, this research provides information about the effects induced by sprouting process on the chemical and physical properties of bran and how these features affect its interaction with water and gluten in both slurry (using the GlutoPeak test) and dough systems. Last but not least, the effect on bread were considered. Unlike the numerous studies present in the literature, where the enzymatic treatments were conducted directly on bran, isolated after milling (i.e., fermentation, addition of enzymes), in this study the bio-technological treatment was performed directly on the wheat kernels. Using bran from sprouted wheat in bread-making has led to positive effects in terms of gluten aggregation kinetic, bread volume and crumb softness, compared to the use of conventional bran. Thus, bran from sprouted wheat might represent a valid strategy to produce staple foods rich in fiber with high quality traits. Moreover, the use of milling by-products is a good approach to decrease food losses. Finally, results from this study might encourage the use of bran from sprouted wheat in small amount in bread formulations to replace the use of exogenous enzymatic improvers and create clean-label products.

**Author Contributions:** Investigation and formal analysis, G.C., P.D., and M.C.C.; Writing—original draft preparation, G.C. and P.D.; Methodology, A.M. and M.C.C.; Conceptualization, writing—review and editing, supervision and project administration, A.M. All authors have read and agreed to the published version of the manuscript.

**Funding:** This research received no external funding.

**Acknowledgments:** The authors wish to thank M.A. Pagani for the critical discussion during the manuscript preparation and Molino Quaglia S.p.A. (Vighizzolo d'Este, Italy) for providing the samples. Microscopy observations were carried out at The Advanced Microscopy Facility Platform—UNItech NOLIMITS—Università degli Studi di Milano (Milan, Italy).

**Conflicts of Interest:** The authors declare no conflict of interest.

## References

1. Sibakov, J.; Lehtinen, P.; Poutanen, K. Cereal brans as dietary fibre ingredients. In *Fibre-Rich and Wholegrain Foods*; Delcour, J.A., Poutanen, K., Eds.; Woodhead Publishing Ltd.: Cambridge, UK, 2013; pp. 170–192. ISBN 9780857090386.
2. Gan, Z.; Galliard, T.; Ellis, P.R.; Angold, R.E.; Vaughan, J.G. Effect of the outer bran layers on the loaf volume of wheat bread. *J. Cereal Sci.* **1992**, *15*, 151–163. [CrossRef]
3. Laurikainen, T.; Härkönen, H.; Autio, K.; Poutanen, K. Effects of enzymes in fibre-enriched baking. *J. Sci. Food Agric.* **1998**, *76*, 239–249. [CrossRef]
4. Pomeranz, Y.; Shogren, M.D.; Finney, K.F.; Bechtel, D.B. Fiber in breadmaking-effects on functional properties. *Cereal Chem.* **1977**, *54*, 25–41.
5. Hemdane, S.; Jacobs, P.J.; Dornez, E.; Verspreet, J.; Delcour, J.A.; Courtin, C.M. Wheat (*Triticum aestivum* L.) bran in bread making: A critical review. *Compr. Rev. Food Sci. Food Saf.* **2016**, *15*, 28–42. [CrossRef]
6. Zhang, D.; Moore, W.R. Wheat bran particle size effects on bread baking performance and quality. *J. Sci. Food Agric.* **1999**, *79*, 805–809. [CrossRef]
7. Marti, A.; Barbiroli, A.; Bonomi, F.; Brutti, A.; Iametti, S.; Marengo, M.; Miriani, M.; Pagani, M.A. Effect of high-pressure processing on the features of wheat milling by-products. *Cereal Chem.* **2014**, *91*, 318–320. [CrossRef]
8. Marti, A.; Bottega, G.; Casiraghi, M.C.; Faoro, F.; Iametti, S.; Pagani, M.A. Dietary fibre enzymatic treatment: A way to improve the rheological properties of high-fibre-enriched dough. *Int. J. Food Sci. Technol.* **2014**, *49*, 305–307. [CrossRef]
9. Katina, K.; Juvonen, R.; Laitila, A.; Flander, L.; Nordlund, E.; Kariluoto, S.; Piironen, V.; Poutanen, K. Fermented wheat bran as a functional ingredient in baking. *Cereal Chem.* **2012**, *89*, 126–134. [CrossRef]
10. Cardone, G.; D'Incecco, P.; Pagani, M.A.; Marti, A. Sprouting improves the bread-making performance of whole wheat flour (*Triticum aestivum* L.). *J. Sci. Food Agric.* **2020**, in press. [CrossRef]
11. Poudel, R.; Finnie, S.; Rose, D.J. Effects of wheat kernel germination time and drying temperature on compositional and end-use properties of the resulting whole wheat flour. *J. Cereal Sci.* **2019**, *86*, 33–40. [CrossRef]
12. Marti, A.; Cardone, G.; Nicolodi, A.; Quaglia, L.; Pagani, M.A. Sprouted wheat as an alternative to conventional flour improvers in bread-making. *LWT Food Sci. Technol.* **2017**, *80*, 230–236. [CrossRef]
13. Marti, A.; Cardone, G.; Pagani, M.A.; Casiraghi, M.C. Flour from sprouted wheat as a new ingredient in bread-making. *LWT Food Sci. Technol.* **2018**, *89*, 237–243. [CrossRef]
14. Faltermaier, A.; Zarnkow, M.; Becker, T.; Gastl, M.; Arendt, E.K. Common wheat (*Triticum aestivum* L.): Evaluating microstructural changes during the malting process by using confocal laser scanning microscopy and scanning electron microscopy. *Eur. Food Res. Technol.* **2015**, *241*, 239–252. [CrossRef]
15. AACC International. *Approved Methods of Analysis*, 11th ed.; AACC International, Inc.: St. Paul, MN, USA, 2011.

16. Zygmunt, L.C.; Anderson, E.; Behrens, B.; Bowers, R.; Bussey, M.; Cohen, G.; Colon, M.; Deis, C.; Given, P.S.; Granade, A. High pressure liquid chromatographic determination of mono-and disaccharides in presweetened cereals: Collaborative study. *J. Assoc. Off. Anal. Chem.* **1982**, *65*, 256–264.
17. *Official Methods of Analysis of AOAC International*, 19th ed.; Official Method 991.43; AOAC International: Gaithersburg, MD, USA, 2012.
18. Manini, F.; Brasca, M.; Plumed-Ferrer, C.; Morandi, S.; Erba, D.; Casiraghi, M.C. Study of the Chemical Changes and Evolution of Microbiota During Sourdoughlike Fermentation of Wheat Bran. *Cereal Chem. J.* **2014**, *91*, 342–349. [CrossRef]
19. Oberleas, D.; Harland, B. Validation of a column liquid chromatographic method for phytate. *J. AOAC Int.* **2007**, *90*, 1635–1638.
20. Lebesi, D.M.; Tzia, C. Staling of cereal bran enriched cakes and the effect of an endoxylanase enzyme on the physicochemical and sensorial characteristics. *J. Food Sci.* **2011**, *76*, S380–S387. [CrossRef]
21. Zanoletti, M.; Marti, A.; Marengo, M.; Iametti, S.; Pagani, M.A.; Renzetti, S. Understanding the influence of buckwheat bran on wheat dough baking performance: Mechanistic insights from molecular and material science approaches. *Food Res. Int.* **2017**, *102*, 728–737. [CrossRef]
22. International Association for Cereal Science and Technology. *Standard No. 115/1, Method for Using the Brabender Farinograph*; International Association for Cereal Science and Technology: Vienna, Austria, 1992.
23. Marti, A.; Torri, L.; Casiraghi, M.C.; Franzetti, L.; Limbo, S.; Morandin, F.; Quaglia, L.; Pagani, M.A. Wheat germ stabilization by heat-treatment or sourdough fermentation: Effects on dough rheology and bread properties. *LWT Food Sci. Technol.* **2014**, *59*, 1100–1106. [CrossRef]
24. Li, Y.; Lu, J.; Gu, G.; Shi, Z.; Mao, Z. Studies on water-extractable arabinoxylans during malting and brewing. *Food Chem.* **2005**, *93*, 33–38. [CrossRef]
25. Hübner, F.; Arendt, E.K. Germination of Cereal Grains as a Way to Improve the Nutritional Value: A Review. *Crit. Rev. Food Sci. Nutr.* **2013**, *53*, 853–861. [CrossRef] [PubMed]
26. Marti, A.; Augst, E.; Cox, S.; Koehler, P. Correlations between gluten aggregation properties defined by the GlutoPeak test and content of quality-related protein fractions of winter wheat flour. *J. Cereal Sci.* **2015**, *66*, 89–95. [CrossRef]
27. Malegori, C.; Grassi, S.; Ohm, J.-B.; Anderson, J.; Marti, A. GlutoPeak profile analysis for wheat classification: Skipping the refinement process. *J. Cereal Sci.* **2018**, *79*, 73–79. [CrossRef]
28. Skendi, A.; Papageorgiou, M.; Biliaderis, C.G. Effect of barley β-glucan molecular size and level on wheat dough rheological properties. *J. Food Eng.* **2009**, *91*, 594–601. [CrossRef]
29. Michniewicz, J.; Biliaderis, G.G.; Bushuk, W. Effect of added pentosans on some properties of wheat bread. *Food Chem.* **1992**, *43*, 251–257. [CrossRef]
30. Enneking, D.; Wink, M. Towards the elimination of anti-nutritional factors in grain legumes. In *Linking Research and Marketing Opportunities for Pulses in the 21st Century*; Springer: New York, NY, USA, 2000; pp. 671–683.
31. Guillon, F.; Champ, M. Structural and physical properties of dietary fibres, and consequences of processing on human physiology. *Food Res. Int.* **2000**, *33*, 233–245. [CrossRef]
32. Nelson, K.; Stojanovska, L.; Vasiljevic, T.; Mathai, M. Germinated grains: A superior whole grain functional food? *Can. J. Physiol. Pharmacol.* **2013**, *91*, 429–441. [CrossRef]
33. Koehler, P.; Hartmann, G.; Wieser, H.; Rychlik, M. Changes of folates, dietary fiber, and proteins in wheat as affected by germination. *J. Agric. Food Chem.* **2007**, *55*, 4678–4683. [CrossRef]
34. Boskov Hansen, H.; Andreasen, M.; Nielsen, M.; Larsen, L.; Knudsen, B.K.; Meyer, A.; Christensen, L.; Hansen, Å. Changes in dietary fibre, phenolic acids and activity of endogenous enzymes during rye bread-making. *Eur. Food Res. Technol.* **2002**, *214*, 33–42. [CrossRef]
35. Shekiro, J.; Kuhn, E.M.; Selig, M.J.; Nagle, N.J.; Decker, S.R.; Elander, R.T. Enzymatic conversion of xylan residues from dilute acid-pretreated corn stover. *Appl. Biochem. Biotechnol.* **2012**, *168*, 421–433. [CrossRef]
36. Jacobs, P.J.; Hemdane, S.; Dornez, E.; Delcour, J.A.; Courtin, C.M. Study of hydration properties of wheat bran as a function of particle size. *Food Chem.* **2015**, *179*, 296–304. [CrossRef] [PubMed]
37. Singh, A.; Kaur, V.; Kaler, R.S.S. A review on dietary fiber in cereals and its characterization. *J. Appl. Nat. Sci.* **2018**, *10*, 1216–1225. [CrossRef]
38. Foschia, M.; Peressini, D.; Sensidoni, A.; Brennan, C.S. The effects of dietary fibre addition on the quality of common cereal products. *J. Cereal Sci.* **2013**, *58*, 216–227. [CrossRef]

39. Lebesi, D.M.; Tzia, C. Use of endoxylanase treated cereal brans for development of dietary fiber enriched cakes. *Innov. Food Sci. Emerg. Technol.* **2012**, *13*, 207–214. [CrossRef]
40. Gruppen, H.; Kormelink, F.J.M.; Voragen, A.G.J. Enzymic Degradation of Water-unextractable Cell Wall Material and Arabinoxylans from Wheat Flour. *J. Cereal Sci.* **1993**, *18*, 129–143. [CrossRef]
41. Katina, K. High-fibre baking. In *Bread Making: Improving Quality*; Cauvain, S.P., Ed.; Woodhead Publishing Ltd.: Cambridge, UK, 2003; pp. 487–499. ISBN 978-1-85573-553-8.
42. Bucsella, B.; Molnár, D.; Harasztos, A.H.; Tömösközi, S. Comparison of the rheological and end-product properties of an industrial aleurone-rich wheat flour, whole grain wheat and rye flour. *J. Cereal Sci.* **2016**, *69*, 40–48. [CrossRef]
43. Noort, M.W.J.; Van Haaster, D.; Hemery, Y.; Schols, H.A.; Hamer, R.J. The effect of particle size of wheat bran fractions on bread quality—Evidence for fibre-protein interactions. *J. Cereal Sci.* **2010**, *52*, 59–64. [CrossRef]
44. Hídvégi, M.; Lásztity, R. Phytic acid content of cereals and legumes and interaction with proteins. *Period. Polytech. Ser. Chem.* **2003**, *46*, 59–64.
45. Van Hung, P.; Hatcher, D.W.; Barker, W. Phenolic acid composition of sprouted wheats by ultra-performance liquid chromatography (UPLC) and their antioxidant activities. *Food Chem.* **2011**, *126*, 1896–1901. [CrossRef]
46. Selinheimo, E.; Kruus, K.; Buchert, J.; Hopia, A.; Autio, K. Effects of laccase, xylanase and their combination on the rheological properties of wheat doughs. *J. Cereal Sci.* **2006**, *43*, 152–159. [CrossRef]
47. Hartikainen, K.; Poutanen, K.; Katina, K. Influence of Bioprocessed Wheat Bran on the Physical and Chemical Properties of Dough and on Wheat Bread Texture. *Cereal Chem. J.* **2014**, *91*, 115–123. [CrossRef]
48. Van Vliet, T. Strain hardening as an indicator of bread-making performance: A review with discussion. *J. Cereal Sci.* **2008**, *48*, 1–9. [CrossRef]
49. Courtin, C.M.; Gelders, G.G.; Delcour, J.A. Use of two endoxylanases with different substrate selectivity for understanding arabinoxylan functionality in wheat flour breadmaking. *Cereal Chem.* **2001**, *78*, 564–571. [CrossRef]
50. Courtin, C.M.; Roelants, A.; Delcour, J.A. Fractionation−reconstitution experiments provide insight into the role of endoxylanases in bread-making. *J. Agric. Food Chem.* **1999**, *47*, 1870–1877. [CrossRef]
51. Rouau, X.; El-Hayek, M.-L.; Moreau, D. Effect of an enzyme preparation containing pentosanases on the bread-making quality of flours in relation to changes in pentosan properties. *J. Cereal Sci.* **1994**, *19*, 259–272. [CrossRef]
52. Ibrahim, Y.; Dappolonia, B.L. Sprouting in hard red spring wheat. *Bak. Dig.* **1979**, *53*, 17–19.
53. Goesaert, H.; Brijs, K.; Veraverbeke, W.S.; Courtin, C.M.; Gebruers, K.; Delcour, J.A. Wheat flour constituents: How they impact bread quality, and how to impact their functionality. *Trends Food Sci. Technol.* **2005**, *16*, 12–30. [CrossRef]
54. Goesaert, H.; Slade, L.; Levine, H.; Delcour, J.A. Amylases and bread firming—An integrated view. *J. Cereal Sci.* **2009**, *50*, 345–352. [CrossRef]
55. De Leyn, I. Functional additives. In *Bakery Products Science and Technology*; Hui, Y.H., Ed.; Blackwell Publishing: Ames, IA, USA, 2006; pp. 233–241.

*Article*

# Chemopreventive Effect of the Germinated Oat and Its Phenolic-AVA Extract in Azoxymethane/Dextran Sulfate Sodium (AOM/DSS) Model of Colon Carcinogenesis in Mice

**Margarita Damazo-Lima [1], Guadalupe Rosas-Pérez [2], Rosalía Reynoso-Camacho [1], Iza F. Pérez-Ramírez [1], Nuria Elizabeth Rocha-Guzmán [3], Ericka A. de los Ríos [4] and Minerva Ramos-Gomez [1,*]**

1 Research and Graduate Studies in Food Science, School of Chemistry, Autonomous University of Queretaro, Cerro de las Campanas S/N, Querétaro 76010, Mexico; dalima06@gmail.com (M.D.-L.); rrcamachomx@yahoo.com.mx (R.R.-C.); iza.perez@uaq.mx (I.F.P.-R.)
2 School of Natural Sciences, Autonomous University of Queretaro, Av. de las Ciencias S/N, Juriquilla, Querétaro 76230, Mexico; gperosas8@gmail.com
3 TECNM/Instituto Tecnológico de Durango, Felipe Pescador 1830 Ote., Durango 34080, Mexico; nuria@itdposgrado-bioquimica.com.mx
4 Institute of Neurobiology, Universidad Nacional Autónoma de México (UNAM), Campus UNAM-Juriquilla, Querétaro 76230, Mexico; eridie9dic@hotmail.com
* Correspondence: minervaramos9297@gmail.com; Tel.: +52-(442)-192-1200 (ext. 5577); Fax: +52-(442)-192-1304

Received: 21 January 2020; Accepted: 6 February 2020; Published: 10 February 2020

**Abstract:** The consumption of fruits, vegetables, nuts, legumes, and whole grains has been associated with a lower risk of colorectal cancer (CRC) due to the content of natural compounds with antioxidant and anticancer activities. The oat (*Avena sativa* L.) is a unique source of avenanthramides (AVAs), among other compounds, with chemopreventive effects. In addition, oat germination has shown enhanced nutraceutical and phytochemical properties. Therefore, our objective was to evaluate the chemopreventive effect of the sprouted oat (SO) and its phenolic-AVA extract (AVA) in azoxymethane (AOM)/dextran sulfate sodium (DSS)-induced CRC mouse model. Turquesa oat seeds were germinated (five days at 25 °C and 60% relative humidity) and, after 16 weeks of administration, animals in the SO- and AVA-treated groups had a significantly lower inflammation grade and tumor (38–50%) and adenocarcinoma (38–63%) incidence compared to those of the AOM+DSS group (80%). Although both treatments normalized colonic GST and NQO1 activities as well as erythrocyte GSH levels, and significantly reduced cecal and colonic β-GA, thus indicating an improvement in the intestinal parameters, the inflammatory states, and the redox states of the animals, SO exerted a superior chemopreventive effect, probably due to the synergistic effects of multiple compounds. Our results indicate that oats retain their biological properties even after the germination process.

**Keywords:** germinated oat; avenanthramides; colorectal cancer; chemoprevention

---

## 1. Introduction

Colorectal cancer (CRC) is the third most diagnosed malignancy and the fourth leading cause of cancer death around the world, and its burden is expected to increase by 60% to more than 2.2 million new cases and 1.1 million cancer deaths between now and 2030 [1]. CRC presents as a series of genetic and morphological changes in the colonic epithelium, which begins with the formation of aberrant crypt foci (ACF), followed by polyps and adenomas, until the development of adenocarcinomas occurs [2]. The etiology of this disease is diverse; however, there are two main risk factors associated: 80–90% of CRC cases are due to environmental factors such as diet and lifestyle, while only 10–20% are due to hereditary factors or genetic alterations. In this regard, diets high in saturated fat and the consumption of processed red meats along with diets low in vegetable and cereal intake increase the risk of CRC [3–5].

Cereals constitute the main source of food worldwide; they are also considered to have high nutritional value, as they contain several beneficial elements, such as starch, proteins, fiber, lipids, and phytochemicals [6]. Oats are an important source of livestock feed worldwide, both as forage and as a nutritious grain. Like all other cereal grains, oats belong to the Pomaceae family (also known as the Gramineae). *Avena sativa* L. (common oat) is considered the most important species among cultivated oats [7]. In this sense, oats of the Turquesa variety, derived from a cross with the Karma variety, have the characteristics of high adaptation, yield stability, and disease-resistance [8].

In recent years, oats have attracted growing attention as a health food, involved in a lower risk of cardiovascular diseases (CVD), type 2 diabetes mellitus (T2DM), gastrointestinal disorders, and cancer [9]. Those properties have been attributed to their content of various bioactive compounds, such as β-glucan, the main component of the soluble fiber in oats. There is evidence revealing that soluble and insoluble β-glucan exerted favorable effects in preventing colon cancer in a dose-dependent manner, although the specific mechanism might be different [10]. Furthermore, oat β-glucan exhibited an anti-inflammatory effect against colitis through inhibition of expression of pro-inflammatory factors in a DSS-induced ulcerative colitis model [11]. In addition, oats have been associated with the presence of several antioxidant compounds, such as tocopherols and tocotrienols, which have the ability to act as strong free-radical scavengers, also showing the capacity to inhibit the proliferation of malignant colon cancer cells [12]. The inhibitory effects of avenacosides (oat-unique steroidal saponins) against the growth of human colon cells has also been evaluated through diverse mechanisms, such as the inhibition of tumor cell growth by cell cycle arrest and the stimulation of apoptosis, among others [9]. Moreover, phenolic compounds, such as ferulic acid and avenanthramides (AVA), are the most abundant components in oats and have documented antioxidant, anti-inflammatory, and antiproliferative activities [13,14]. In vitro studies indicate that AVAs prevent cancer mainly by blocking reactive species, and exhibit potential therapeutic activity through the modulation of different pathways, including the activation of apoptosis and senescence and the blocking of cell proliferation. In this context, AVA-A (2p) notably attenuated tumor formation in an azoxymethane (AOM)/dextran sulfate sodium (DSS) model, most likely through the induction of cellular senescence [15].

On the other hand, cereal sprouts have received considerable attention as functional foods in many countries, especially in Europe, the United States of America, and Japan, due to their superior nutritious and health benefits compared to seeds [16]. The germination of cereals has been used for centuries to soften kernel structure, decrease the content of antinutritive compounds, and to increase nutrient content and availability [17]. In this regard, an increase in the nutraceutical and phytochemical profile of oats could improve their biological activities, such as in the case of AVA-C (2c) from germinated oats, which exhibited antiallergic and anti-inflammatory activities [18]. Additionally, numerous trials and animal studies have demonstrated the bioavailability and bioaccessibility of polyphenolic compounds of oats, especially AVAs, and their association with an improved antioxidant status [19–22]. Similarly, mounting evidence highlights that the absorption, bioavailability, and metabolism of several phytochemicals is a crucial factor in determining their biological activity against colon cancer [23]. Therefore, the metabolism of dietary flavonoids in the digestive tract by the gut microbiota is important for increasing their bioavailability and determining their impact on our health.

To the best of our knowledge, only a few studies have focused on the relationship between an enhanced phytochemical profile and the biological values of cereal sprouts, and there is currently not enough available information regarding the chemopreventive activity of germinated oats. Therefore, the purpose of this investigation was to evaluate whether the sprouted oat (SO) of the Turquesa variety still possessed effective physiologically bioactive compounds, i.e., phenolics, flavonoids, AVAs, and phytosterols, and whether it exerted antioxidant and anti-inflammatory effects, as well as the capacity to improve relevant intestinal parameters (pH and β-GA activities) in an AOM/DSS-induced CRC mouse model.

## 2. Materials and Methods

### 2.1. Oat Seed Germination

Oat (*Avena sativa* L.) seeds of the "Turquesa" variety were donated by Instituto Nacional de Investigaciones Forestales, Agrícolas y Pecuarias (INIFAP) Campo Experimental Bajío, Celaya, GT, MEX. For the germination process, seeds (100 g) were soaked in 1.5% sodium hypochlorite (1:6 *w*/*v*) for 30 min at 30 °C. Then, seeds were drained and washed with distilled water until they reached a neutral pH. Afterwards, seeds were soaked in distilled water (1:6 *w*/*v*) for 12 h with occasional shaking. Finally, hydrated seeds were placed in trays where a wet filter paper was extended, and then covered. The trays were introduced into a germination chamber, and filter paper was watered daily; chamber temperature and relative humidity (RH) were set at 25 °C and 60% RH, respectively. Germination was performed in darkness for five days and the experiment included three replicates. Based on previous studies, these conditions proved to be effective in ensuring the germination of the oat seeds. At the end of the process, the germination percentage was determined based on the total number of fully emerged seedlings, and the radicle length (mm) was measured using a Vernier caliper. Afterwards, oat sprouts were manually cleaned up from impurities and soil contaminants, and were immediately dried at 50 °C for 12 h, ground in a mill, and passed through a mesh with a particle size of 0.5 mm. Finally, the sprouted oat (SO) flours were stored at 4 °C until the analyses.

### 2.2. Chemical Analysis (Proximate Analysis) and Phytic Acid Content

The Official Methods of Analysis of the Association of Official Agricultural Chemists (AOAC) were used to determine the moisture (method 925.10), crude fat (method 920.85), protein (method 920.87), ash (method 923.03) [24], and crude fiber [25] contents of SO flours. The carbohydrate content was calculated by difference. The insoluble and soluble fiber contents were determined by a Megazyme β–Glucan Assay Kit Mixed Linkage (Megazyme kit, Wicklow, IRL). Insoluble and soluble fiber contents are expressed as a percentage. The phytic acid content was determined according to the method of Latta and Eskin [26] and expressed as a percentage. All determinations were performed according to three independent experiments, which included mean values of three technical repetitions.

### 2.3. Methanolic Extraction and Quantification of Total Phenolic Compounds (TPCs)

TPCs were quantified in the SO flour samples according to Singleton and collaborators [27]. Briefly, extractions with methanol/water (50:50, *v*/*v*) acidified with HCl (pH 2) and acetone/water (70:30, *v*/*v*) were combined and used for the quantification of TPC using the Folin–Ciocalteu procedure. The results of the TPCs were expressed as mg of gallic acid equivalents (GAE)/g of SO flour.

### 2.4. Abundance of Phenolic Compounds by Ultra–Performance Liquid Chromatograph (UPLC)–Mass Spectrometre (MS)

The polyphenolic extract previously mentioned was concentrated and assessed by an ultra–performance liquid chromatograph (UPLC) coupled to a photodiode array (PDA) detector and a quadrupole time-of-flight (QTOF)–MS with an atmospheric pressure electrospray ionization (ESI) (Vion, Waters Co., Milford, MA, USA) interphase for the identification of the compositions of polyphenols and AVAs. The polyphenol extract was resuspended in 1 mL of the mobile phase (95% acidified water with 0.1% formic acid and 5% of acetonitrile acidified with 0.11% formic acid) and filtered (Polyvinylidene fluoride syringe filters, 0.2 μm). Samples were kept at 10 °C in the automated injector system and injected (1 μL) into an Acquity BEH column (100 × 2.1 mm, 1.7 μm) at 35 °C under gradient conditions. The ESI–QTOF–MS conditions were as follows: Data acquisition, negative ionization mode (ESI-); mass range, 100–1200 Da; capillary voltage, 2.0 kV; cone voltage, 40 eV; collision energy, 6 V (low) and 15–45 V (high); source temperature, 120 °C; desolvation gas ($N_2$), 450 °C at 800 L/h; cone gas flow, 50 L/h; lock mass correction, leucine enkephalin (50 pg/mL) at 10 µL/min.

UPLC-QTOF analysis was made as previously reported by Rodríguez–González and collaborators [28]. Peak identity was established by comparison of the exact mass of the pseudomolecular ion (confirmation through molecular composition with a mass error <10 ppm, isotopic distribution, and fragmentation pattern). Mass spectra were analyzed in comparison with the following libraries: Phenol-Explorer, Food Database, and PubChem.

### 2.5. Animals and Treatments

Male CD-1 mice (3–4 weeks of age) were purchased from the Institute of Neurobiology, Campus UNAM-Juriquilla (QT, MEX). The experiments on animals were performed in accordance with the Animal Care and Use protocol and were approved by the Ethics Committee of the Autonomous University of Queretaro (Project identification code: CBQ18/069, approved: 19 June 2018). Animals were housed in metallic cages (two or three mice per cage) and maintained under controlled conditions (12 h dark/light cycles at 25 °C and 50% ± 10% RH). Mice had free access to water and a standard pellet diet (rodent diet 501, LabDiet, USA). After two weeks of acclimatization, mice were randomly assigned to four experimental groups. Group 1: Normal ($n$ = 10); group 2: AOM + DSS as control group ($n$ = 10); group 3: AVA + AOM + DSS ($n$ = 8); group 4: SO + AOM + DSS ($n$ = 8). Animals in groups 1 and 2 were intragastrically fed with a vehicle (200 µL of saline solution). Mice in groups 3 and 4 were gavaged every morning with the phenolic-AVA extract (0.084 mg GAE/day) and 30 mg/day of SO (equivalent to 0.084 mg of phenolic extract), respectively, dissolved in 200 µL of saline solution as a vehicle, throughout the 16 week period, with an exception at week 6. After four weeks, animals in groups 2–4 were intraperitoneally injected with AOM (10 mg/kg; dissolved in NaCl 0.9%); a week later, animals received DSS (2% in the drinking water) ad libitum for seven days. Mice were weighed weekly; stool samples of 24 h were collected at week 15. Animals in all groups were euthanized by guillotine decapitation after a 16 week experimental period. Immediately after decapitation, the blood of each animal was collected, mixed by inversion to prevent clot formation, and centrifuged. The plasma and erythrocytes were separated, and erythrocytes were lysed. Afterward, the colons of the rats were immediately excised, and cecal and colonic contents were collected separately, immediately frozen, and stored at −70 °C for pH and β-GA analyses. The livers were also removed, cleansed with saline solution, and frozen with liquid nitrogen. All samples and organs were stored at −70 °C until their utilization for the corresponding analysis.

### 2.6. Assay of β-Glucuronidase Activity (β-GA) and pH in the Cecal, Colonic, and Fecal Contents

Cecal and colonic contents, as well as feces, were suspended in high–performance liquid chromatograph (HPLC) water (relation 1:5 *w/v*) and sonicated with icing for 3 min at 4 °C. Samples were centrifugated at 500 g for 15 min at 4 °C, and the supernatants were used for β-GA and pH values according to Jenab and Thompson [29]. β-GA was measured at 540 nm, and the amount of phenolphthalein released was determined using a phenolphthalein standard curve. β-GA is expressed as µg phenolphthalein per hour per g content.

### 2.7. Macroscopic and Histopathology Analyses

Colons were rinsed with saline solution, opened longitudinally, divided into proximal and distal sections, and inspected for macroscopic pathological lesions. Macroscopic lesions were cut and fixed in 10% buffered formalin, embedded in paraffin blocks, and processed for subsequent hematoxylin and eosin (H&E) staining and tumor classification according to Astler and Coller [30]. Lesions of the colon were classified as normal, inflammations (grade: +, ++, and +++), or adenocarcinomas [31]. The histopathological analysis of colonic lesions also consisted of the classification of infiltrating lymphocytes, presence of eosinophils, calceiform cells, muscularis externa, necrosis area, and mitosis percentage as a strategy to further characterize colonic inflammation in all groups [31–34]. The histopathological examination was performed by two observers independently. Differences of one or more grades were re-examined by both observers before reaching a final consensus.

### 2.8. Erythrocyte-Reduced Glutathione (GSH) Level

GSH content in erythrocytes was measured according to Ellman's procedure [35]. First, erythrocytes were lysed and sulfosalicylic acid (5%) was added as a protein precipitation agent; then, samples were centrifuged, and the supernatants were assayed for GSH content. Protein concentration was determined by the bicinchoninic acid (BCA) protein assay kit (Pierce Thermo Scientific) using bovine serum albumin (BSA) as the standard. Levels of GSH are expressed as µM of GSH per mg protein.

### 2.9. Phase 2 Enzyme Assays

Total glutathione S-transferase (GST) and NAD(P)H:quinone oxidoreductase 1 (NQO1) activities in colonic and hepatic cytosolic fractions were measured according to the methods by Habig and collaborators [36] and Prochaska and collaborators [37], respectively. Protein concentration was determined by the BCA assay using BSA as standard. Enzymatic activities are expressed as nmol of product per min per mg protein.

### 2.10. Statistical Analysis

Results are expressed as means ± standard error (SE), except for the chemical and nutraceutical characterization of SO and oat seed data, which are expressed as means ± standard deviation (SD). Statistical significance for chemical composition and nutraceutical compounds of SO was determined by Student's test at $p < 0.05$. For the in vivo study, statistical significance was determined by ANOVA followed by Tukey's test at $p < 0.05$. The Chi-square test was used to determine differences in the macroscopic and histopathological classifications of colonic lesions. Correlations were assessed by Pearson correlation analysis at $p < 0.05$. Statistical analyses were performed using JMP version 11.0.0 (Systat Software, Inc., San José, CA, USA).

## 3. Results

### 3.1. Characterization of Chemical Composition and Nutraceutical Compounds of Oat Seeds after the Germination Process

An increase in phenolic compounds, as well as changes in chemical composition, was demonstrated during the germination of grains. Therefore, the proximal compositions and nutraceutical determinations of SO flour and oat seeds are shown in Table 1; in addition, a measurement of phytic acid was carried out after five days of germination in darkness at 25 °C/60% RH. Under this condition, we reached 100% of germination and a radicle length of 6.47 ± 0.22 cm. Protein and lipid contents, as well as moisture, were higher in SO, whereas carbohydrate and ash contents were lower after oat seed germination at 25 °C/60% RH. Similarly, total fiber was also lower in SO, and soluble fiber was most affected by the germination process; more importantly, phytic acid, an antinutritional compound present in oats, was 10 times lower in oat sprouts.

As expected, TPC was higher in the germination condition (25 °C/60% RH) compared with that of the raw seed. Therefore, we aimed to identify some of the phenolic families present in the methanolic extract of SO by UPLC-MS.

**Table 1.** Chemical analysis, insoluble and soluble dietary fiber, phytic acid, and total phenolic contents in oat seed and SO (Turquesa variety) after five days of germination in darkness at 25 °C/60% relative humidity (RH).

| Sample | Protein [1] | Carbohydrates [2] | Lipids [1] | Ash [1] | Moisture [1] |
|---|---|---|---|---|---|
| Oat seed | 8.83 ± 0.23 [a] | 73.37 ± 1.1 [a] | 4.41 ± 0.20 [a] | 3.99 ± 0.30 [a] | 9.4 ± 0.7 [a] |
| SO | 10.05 ± 0.30 [b] | 71.08 ± 0.7 [b] | 5.55 ± 0.01 [b] | 3.62 ± 0.10 [b] | 9.7 ± 0.2 [a] |
| | **Insoluble Fiber [1]** | **Soluble Fiber [1]** | **Total Fiber [1]** | **Phytic Acid [1]** | **TPC [3]** |
| Oat seed | 55.04 ± 1.89 [a] | 18.49 ± 0.74 [a] | 73.53 ± 2.51 [a] | 0.44 ± 0.0 [a] | 0.64 ± 0.01 [a] |
| SO | 30.87 ± 4.23 [b] | 2.28 ± 0.52 [b] | 33.15 ± 4.65 [b] | 0.04 ± 0.0 [b] | 2.79 ± 0.06 [b] |

The results represent the average of three replicates ± standard deviation (SD). [1] Expressed as percentage (%) on dry basis, except for moisture. [2] Calculated by difference. [3] Expressed as mg of gallic acid equivalent (GAE)/g. TPC: Total phenolic compounds. SO: Sprouted oat. Values with different letter(s) within a column are significantly different according to Student's test ($p < 0.05$).

### 3.2. *Phytochemical Profile Induced after Germination of Oat Seeds at 25 °C/60% RH*

This study identified polyphenolic families previously reported in the literature, such as AVAs, saponins, hydroxybenzoic acids, flavonols, and flavones. In addition, families not previously reported in the literature in oat grains and SO were identified here, which include mainly isoflavones, lignans, and phytosterols.

Table S1 (Supplementary Material) shows the phytochemical profiles of oat seeds and SO. Firstly, SO exhibited four major AVAs: AVA-D, AVA-L, AVA-G/1c/2p isomer I, and AVA-B, with AVA-D being the most abundant. As for the family of saponins, avenacoside A was identified in both whole oats and germinated seeds. (Epi)-catechin isomer II and (Epi)-catechin hexose isomer II were the major flavanols in SO. On the other hand, apigenin apiosyl-hexoside and luteolin apiosyl-malonyl-hexoside were identified as the two major flavones, whereas low levels of apigenin glucuronide, luteolin 7-O-hexoside, and hydroxyluteolin were detected in SO. Kaempferol pentoside-hexoside-rhamnoside was the flavonol most abundantly detected in SO, whereas quercetin dihexoside, quercetin xyloside, kaempferol, and quercetin acetyl-hexoside-rhamnoside were present with lower abundance.

We identified members of the hydroxybenzoic acids family in raw seeds and SO that was not previously reported in the literature, such as gallic acid hexoside isomer I, hydroxybenzoic acid isomer II, and benzoic acid, while protocatechuic acid was only detected in SO. Acetylgenistin and acetyldaidzin were the two main isoflavones in SO, whereas hydroxydihydrodaidzein isomer II was only detected in SO with the lowest abundance. Acetoxypinoresinol and secoisolariciresinol were identified as the major lignans. It is important to note that there are no previous reports in which the lignin family has been identified in oat seeds. On the other hand, the most abundant phytosterols were β-campesterol hexoside and β-sitosterol hexoside; conversely, β-campesterol was only detected in SO with the lowest abundance.

### 3.3. *Effect of Sprouted Oat (SO) and Its Phenolic AVA Extract (AVA) on Body Weight and Intestinal Parameters in CD–1 Mice Induced with AOM and DSS*

In order to evaluate the chemopreventive efficacy of the sprouted oat (SO) and its phenolic AVA extract (AVA) in an AOM/DSS model, body weight was monitored weekly and pH values and β-GA activity were determined in cecal, colonic, and fecal samples (Table 2). At the end of the 16 week experimental period, the Normal and AOM + DSS-treated groups showed similar body weight gains. Therefore, no statistical differences were found in the final weights among the experimental groups ($p > 0.05$). These results suggest that, under experimental conditions, the carcinogenic and promoting agents (AOM + DSS), as well as the SO and AVA treatments, did not cause adverse effects related to nutrient absorption or animal growth. However, we observed anal discomfort in some mice from the 14th week of induction with AOM + DSS. In the AOM + DSS control group, four animals had rectal bleeding and five animals developed anal prolapses; meanwhile, only two animals from the AVA group and three from the SO group developed anal bleeding, and only one from each group developed anal prolapse. From week 15, bloody feces and diarrhea were more frequently observed in the AOM + DSS control group compared to those in AVA- and SO-treated groups. Animals with these conditions were individually assigned in cages and evaluated by a veterinarian on a routine basis. Regarding the survival of the animals under study, Normal and AOM + DSS control groups showed a survival rate of 100%, while AVA- and SO-treated groups showed an 89% survival rate with the death of one animal in each group, which, according to the autopsies performed by the veterinarian, were due to causes other than cancer (i.e., lack of the development of the frontal incisors and bodily injury in the metabolic cage).

As we previously reported, increased β-GA and pH values have been related to the carcinogenic effect of the AOM in the presence of DSS [38]. Therefore, β-GA and pH values in fecal, cecal, and colonic samples collected before and during the euthanizing of the animals were also determined, and the results are shown in Table 2. Although not statistically different, all AOM + DSS-treated groups had higher pH values in cecal, colonic, and fecal samples, with the values of the SO + AOM + DSS group most similar to those of the Normal group. As expected, the level of β-GA in the cecal, colonic, and fecal samples of the AOM + DSS control group was statistically higher than those of the Normal group ($p < 0.05$). Interestingly, both AVA and SO treatments significantly reduced β-GA levels in cecal and colonic samples, with the highest effect occurring in colonic β-GA values of the SO-treated groups (36% lower in comparison with that of the AOM + DSS control group).

### *3.4. Anticarcinogenic Effect of Sprouted Oat (SO) and Its Phenolic AVA Extract (AVA) on the Macroscopic and Histopathological Quantitative Classification of Colonic Lesions Induced with AOM and DSS in Male CD–1 Mice*

The lesions found in the colon of each of the animals were classified macroscopically as flat-type lesions, called early lesions or plaques, and more advanced lesions, called polyps or tumors. The latter are defined as protuberances of cells that protrude into the intestinal lumen, which are characterized by increased cell division, and can be benign and asymptomatic lesions or capable of evolving into malignant lesions [39,40].

Table 3 shows the results derived from the early lesion and tumor counts found in mice of the experimental AOM + DSS model. The Normal group had an early lesion incidence of 10%, and no tumors were developed in animals in this group ($p < 0.05$). On the other hand, AVA- and SO-treated groups had the highest incidence of early lesions (100%) compared to that of the AOM + DSS control group (60%); however, the AOM + DSS control group showed the highest tumor incidence (80%) of all AOM + DSS-treated groups.

In addition, the distribution of the lesions (flat-type lesions and tumors) was also analyzed. As expected, most of the plaques and tumors in AOM + DSS-treated groups were found in the distal portion of the colon (80–100%) and, according to tumor classification, polyps in the AOM + DSS control group were mostly of the sessile type, followed by pedunculate-type polyps. Interestingly, both tumor types were reduced in SO- and AVA-treated groups; moreover, animals in the SO-treated group developed only pedunculate polyps.

The histopathological study also included analyses of the inflammation grade and dysplasia, as well as the incidence of adenocarcinomas; however, only inflammation grade and adenocarcinomas were considered (Table 4, Figure 1). The Chi-square test revealed a significant difference ($\alpha = 0.05$) with respect to the incidence of adenocarcinomas. As expected, none of the animals in the Normal group developed adenocarcinomas; however, one of them (10%) presented a medium inflammation grade (++). In the AOM + DSS control group, 80% of the animals developed adenocarcinomas, and 20% showed an inflammation grade of +++. Mice in the SO + AOM + DSS group exhibited the lowest adenocarcinoma and inflammation incidence (38%); however, the phenolic AVA extract did not exert the same protection, with the incidences of adenocarcinomas and inflammation being 63% and 25%, respectively.

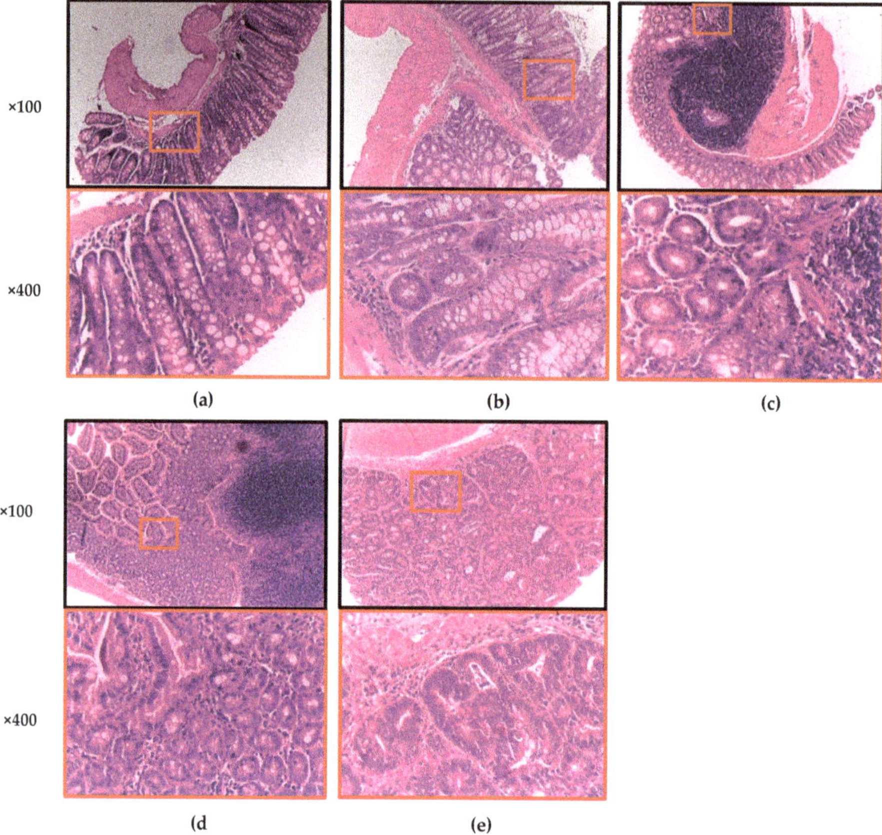


**Figure 1.** Histopathological examination of colonic mucosa of azoxymethane (AOM) + dextran sulfate sodium (DSS)-treated mice. Hematoxylin and eosin (H&E) staining: (**a**) Normal tissue; (**b**) low-grade inflammation (+); (**c**) medium-grade inflammation (++); (**d**) high-grade inflammation (+++); (**e**) adenocarcinoma. $n$ = 8–10 mice per group.

As inflammation was still present at the end of the experimental period (16 weeks), a quantitative and descriptive classification of the damage in the colon based on lymphocyte infiltration, eosinophils, calceiform cells, epithelial ridges, necrosis, and mitosis presence in tissues of the colon was carried out as a strategy to characterize colonic inflammation (Table 4, Figure 2). In this sense, AVA- and SO-treated groups showed a similar behavior: Medium lymphocyte infiltration, the presence of eosinophils and epithelial ridges, few calceiform cells, an absence of necrosis, and <20% of mitosis. Conversely, the AOM + DSS group showed high lymphocyte infiltration, the presence of eosinophils and necrosis, fewer calceiform cells, an absence of epithelial ridges, and 20% of mitosis.

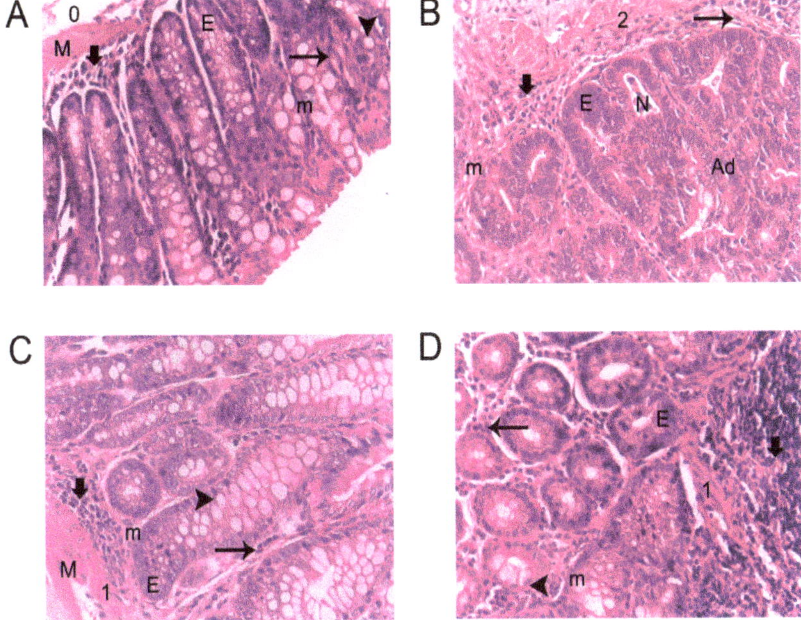


**Figure 2.** Histopathological features of colonic mucosa of AOM + DSS-treated mice. H&E staining: (**A**) Normal group, (**B**) AOM + DSS group, (**C**) AVA + AOM + DSS group, (**D**) SO + AOM + DSS group. $n = 8–10$ mice per group. (→) Lymphocyte infiltration; (⬇) eosinophils; (➤) goblet cells; (E) epithelial ridges; (M) defined muscularis; (m) mitosis; (N) necrosis; (0,1,2) inflammation grade; and (Ad) adenocarcinoma.

### 3.5. Antioxidant Effects of Sprouted Oat (SO) and Its Phenolic AVA Extract (AVA) in the AOM/DSS Model

A widely accepted mechanism in cancer chemoprevention by dietary phytochemicals is through the induction of antioxidant and cytoprotective systems, such as reduced glutathione (GSH), and the activity of the phase 2 enzymes GST and NQO1, among others, through the activation of Nrf2 (NF-E2-related factor 2) signaling pathways [41]. Regarding the serum concentration of GSH, the level of GSH was significantly increased (twofold) in animals induced with AOM + DSS compared to that of animals in the Normal group (Table 5). It is important to highlight that SO and AVA treatments normalized serum GSH levels in the animals of both groups. In addition, hepatic and colonic GST activities were significantly higher in the AOM + DSS control group compared to those in the Normal group. Similarly, AVA and SO treatments achieved normalization of GST and NQO1 activities in the colon, whereas the SO + AOM + DSS group had the highest hepatic GST and NQO1 activities.

**Table 2.** Changes in body weight and cecal, colonic, and fecal measurements of CD-1 mice fed with sprouted oat (SO) and its phenolic AVA extract (AVA) at week 16 of the AOM + DSS-induced colon carcinogenesis.

| | Body Weight | | Cecum | | Colon | | Feces | |
|---|---|---|---|---|---|---|---|---|
| Group | Initial | Final | pH | β-GA [1] | pH | β-GA [1] | pH | β-GA [1] |
| Normal | 30.6 ± 1.01 $^{a}$ | 39.8 ± 1.01 $^{a}$ | 7.12 ± 0.04 $^{b}$ | 3.67 ± 0.13 $^{c}$ | 7.22 ± 0.07 $^{a}$ | 3.99 ± 0.09 $^{d}$ | 7.22 ± 0.13 $^{b}$ | 4.34 ± 0.17 $^{b}$ |
| AOM + DSS control | 30.0 ± 0.76 $^{a}$ | 41.8 ± 0.79 $^{a}$ | 7.29 ± 0.04 $^{a,b}$ | 5.28 ± 0.26 $^{a}$ | 7.33 ± 0.04 $^{a}$ | 7.02 ± 0.15 $^{a}$ | 7.77 ± 0.12 $^{a}$ | 5.70 ± 0.15 $^{a}$ |
| AVA + AOM + DSS | 28.8 ± 1.03 $^{a}$ | 40.1 ± 1.10 $^{a}$ | 7.46 ± 0.05 $^{a}$ | 3.93 ± 0.08 $^{b,c}$ | 7.31 ± 0.10 $^{a}$ | 6.17 ± 0.06 $^{b}$ | 7.67 ± 0.11 $^{a}$ | 5.87 ± 0.27 $^{a}$ |
| SO + AOM + DSS | 28.9 ± 0.88 $^{a}$ | 38.9 ± 1.03 $^{a}$ | 7.13 ± 0.07 $^{b}$ | 4.11 ± 0.19 $^{b}$ | 7.09 ± 0.12 $^{a}$ | 4.50 ± 0.06 $^{c}$ | 7.62 ± 0.22 $^{a}$ | 2.51 ± 0.20 $^{c}$ |

Values are mean ± standard error, $n$ = 5–10. [1] mg phenolphthalein per hour per g cecal, colonic, and fecal content. Values with different letter(s) within a column are significantly different according to the Tukey test ($p < 0.05$). AVA: phenolic avenanthramide (AVA) extract (0.084 mg GAE/day); SO: sprouted oat (30 mg/day).

**Table 3.** Anticarcinogenic effect of sprouted oat (SO) and its phenolic AVA extract (AVA) on the macroscopic quantitative classification of colonic lesions induced with AOM + DSS in CD-1 mice.

| | | Early Lesions (Flat-Type Lesions) [1] | | | | Tumors (Protuberant-Type Lesions) [2] | | | | | | | |
|---|---|---|---|---|---|---|---|---|---|---|---|---|---|
| Group | $n$ | Incidence (%) | Mean Number [3] | Colon Distribution [4] Proximal | Distal | Incidence (%) | Mean Number [3] | Colon Distribution [4] Proximal | Distal | Classification [5] P | S | EX | EN |
| Normal | 10 | 10 * | 0.10 ± 0.10 $^{b}$ | 1 (100%) | 0 (0%) | 0 * | 0.00 ± 0.00 $^{b}$ | 0 (0%) | 0 (0%) | 0 | 0 | 0 | 0 |
| AOM + DSS | 10 | 60 | 1.00 ± 0.30 $^{a,b}$ | 0 (0%) | 10 (100%) | 80 | 4.20 ± 1.01 $^{a}$ | 0 (0%) | 42 (100%) | 14 | 16 | 2 | 10 |
| AVA + AOM + DSS | 8 | 100 | 2.13 ± 0.30 $^{a}$ | 0 (0%) | 17 (100%) | 50 | 0.63 ± 0.26 $^{b}$ | 1 (20%) | 4 (80%) | 4 | 1 | 0 | 0 |
| SO + AOM + DSS | 8 | 100 | 2.13 ± 0.61 $^{a}$ | 1 (6%) | 16 (94%) | 38 | 0.38 ± 0.18 $^{b}$ | 0 (0%) | 3 (100%) | 3 | 0 | 0 | 0 |

[1] Macroscopic quantitative evaluation of flat-type lesions. [2] Macroscopic quantitative evaluation of protuberant lesions or tumors. [3] Total number of early lesions or tumors/mice per group. [4] Total number of early lesions or tumors and incidence between parenthesis. [5] Tumor classification: P, pedunculate; S, sessile; EX, exophytic, and EN, endophytic. * Incidence is statistically significant according to the Chi-square test ($p < 0.05$). AVA: phenolic AVA extract (0.084 mg GAE/day); SO: sprouted oat (30 mg/day).

**Table 4.** Quantitative classification of adenocarcinomas and histopathological inflammatory features in the AOM + DSS-induced carcinogenic study.

| Group | *n* | Incidence (%) | | | Lymphocyte Infiltration [1] | | | Eosinophils | | Calceiform Cells [2] | | | Epithelial Ridges | | Defined Muscularis | | Necrosis | | Mitosis | Inflammation Grade | | |
|---|---|---|---|---|---|---|---|---|---|---|---|---|---|---|---|---|---|---|---|---|---|---|
| | | Normal Tissue | Inflammation | Adeno Carcinomas | L | M | H | Yes | No | F | M | I | Yes | No | Yes | No | Yes | No | | + | ++ | +++ |
| **Normal** | 10 | 9 (90%) * | 1 (10%) * | 0 (0%) * | | None | | X | | | | X | X | | X | | | X | 5% | | 1 | |
| **AOM + DSS** | 10 | 0 (0%) | 2 (20%) | 8 (80%) | | | X | X | | X | | | | X | | X | X | | 20% | 0 | 0 | 2 |
| **AVA + AOM + DSS** | 8 | 1 (12%) | 2 (25%) | 5 (63%) | | X | | X | | X | | | X | | Focal | | | X | 15% | 1 | 1 | |
| **SO + AOM + DSS** | 8 | 2 (25%) | 3 (38%) | 3 (38%) * | | X | | X | | X | | | X | | X | | | X | 17% | 1 | 2 | |

Histopathological quantitative classification according to hematoxylin and eosin (H&E) staining in colonic tissues (×400). Values are mean ($n$ = 8–10 animals per group). AVA: phenolic AVA extract (0.084 mg GAE/day); SO: sprouted oat (30 mg/day). * Incidence is statistically significant according to the Chi-square test ($p < 0.05$). [1] Lymphocyte infiltration levels, L: Low, M: Medium, H: High. [2] Calceiform cell classification: F: Few, M: Moderate, I: Intense.

**Table 5.** Antioxidant effect of sprouted oat (SO) and its phenolic AVA extract (AVA) in the AOM + DSS-induced colon cancer model.

| | | Erythrocyte | Liver | | Colon | |
|---|---|---|---|---|---|---|
| **Group** | **n** | **GSH [1]** | **GST [2]** | **NQO1 [2]** | **GST [2]** | **NQO1 [2]** |
| **Normal** | 10 | 1.40 ± 0.10 $^{b}$ | 411.5 ± 7.7 $^{a}$ | 36.2 ± 1.0 $^{a,b}$ | 60.6 ± 3.0 $^{a}$ | 54.6 ± 2.2 $^{ab}$ |
| **AOM + DSS** | 10 | 2.81 ± 0.43 $^{a}$ | 438.6 ± 12.3 $^{b}$ | 38.6 ± 0.9 $^{a}$ | 99.2 ± 6.6 $^{b}$ | 88.7 ± 2.3 $^{c}$ |
| **AVA + AOM + DSS** | 8 | 1.39 ± 0.15 $^{b}$ | 417.7 ± 10.0 $^{a}$ | 33.4 ± 0.8 $^{b}$ | 61.9 ± 2.1 $^{a}$ | 47.5 ± 3.3 $^{a}$ |
| **SO + AOM + DSS** | 8 | 1.34 ± 0.24 $^{b}$ | 506.3 ± 2.7 $^{c}$ | 58.7 ± 1.8 $^{c}$ | 55.8 ± 1.6 $^{a}$ | 58.4 ± 1.7 $^{b}$ |

Values are mean ± standard error, $n$ = 5–10. [1] nmol per mg protein. [2] nmol product per min per mg protein. Values with different letter(s) within a column are significantly different according to the Tukey test ($p < 0.05$). AVA: phenolic AVA extract (0.084 mg GAE/day); SO: sprouted oat (30 mg/day).

## 4. Discussion

Numerous epidemiological studies have demonstrated the relationship between high consumption of whole grains (90 g or three servings per day) and the reduced risk of coronary artery disease, cardiovascular disease, total cancer, and all-cause mortality [42]. Similarly, the suboptimal intake of whole grains (38 g/d) was associated with CRC burden across 16 European countries [43]. Therefore, an optimal intake of 50–100 g/d was considered in our study to establish the dose administered in the AOM/DSS-induced CRC mouse model (75 g/d). In addition, the germination of cereal seeds, such as oats, is a technique that has been used for centuries to soften the structure of the grain, improve its nutritional value, and reduce the antinutritional effects, while enhancing the phenolic profile with biological activity [17,44,45]. In this regard, five days of germination in darkness at 25 °C/60% RH proved to be effective in ensuring the germination of Turquesa oat seeds. Under these conditions, changes in chemical composition were expected, since total protein and lipid contents were slightly increased during a five-day germination period due to their remobilization to the developing embryo [45,46]; however, carbohydrate degradation is very limited during germination, despite the fact that starch-degrading enzymes are synthesized [47], which agrees with our results. Dietary fiber, with roughly 60% as insoluble fiber and 40% as soluble fiber [48], also decreases significantly during germination, since the release of gibberellin, a hormone capable of activating the enzyme β-glucanase, promotes β-glucan degradation, among other polysaccharides [17,49], which would explain the decrease in soluble fiber in our sample. Oats also contain other compounds such as phytic acid (5.6–8.7 mg/g; 0.56–0.87%) [9], which has its content decreased by 15%–35% during even a short three-day germination due to activation of phytase activity [50]. Although high doses of phytic acid inhibit the absorption of metals and minerals in humans, it has been observed that, in small doses, it can function as a protective factor in several chronic degenerative diseases [51]; therefore, current research only seeks to reduce the content of this antinutritional compound in various cereals.

Since part of our objective was to understand the chemopreventive effect of oats following oral administration of sprouted oat (SO) or its phenolic AVA extract (AVA), increased TPC after germination was an important keynote for determining whether the rich-phenolic extract mediated the health benefits of oats. In this regard, a more than four-fold increase in TPC was obtained after five days of germination in darkness at 25 °C/60% RH. Similar results have been previously reported [17,45,52]. This result could also reflect the better extractability of phenolic compounds from kernel structures after germination [17].

As we mentioned before, we aimed to identify the families of polyphenols contained in our sprouted oat flour from the Turquesa variety in order to inquire about the reported effects of these compounds, in addition to identifying compounds not previously reported in other oat varieties. Studies about the phytochemical profiles of oat sprouts are limited in comparison with those for whole seeds and hulks. The phenolic compounds reported in oats include phenolic acids, flavonoids, and AVA [53], with the most abundant being N-(3′,4′-dihydroxy-(E)-cinnamoyl)-5-hydroxyanthranilic acid (AVA-C or 2c), N-(4′-hydroxy-3′-methoxy-(E)-cinnamoyl)-5-hydroxyanthranilic acid (AVA-B or 2f), and N-(4′-hydroxy-(E)-cinnamoyl)-5-hydroxyanthranilic acid (AVA-A or 2p). In this study, we identified AVA-D as the most abundant AVA, followed by AVA-L, which had not been reported as one of the three most abundant AVAs in other oat varieties. In addition, of the three most abundant AVAs previously reported, only AVA-B had a higher abundance in the germination condition, confirming previous reports that indicate that AVA concentration depends on variety, fraction, genotype, and the growing environment conditions [54–56]. Generally, sprouts contain measurable amounts of flavonoids such as catechin and epicatechin [57]. In this regard, we identified some isomers in SO; (Epi)-catechin isomer II was the major flavanol. Other individual phenolic compounds present in SO were organic acids (hydroxybenzoic and hydroxycinnamic); the ellagic acid was detected in major abundance in SO, followed by benzoic, hydroxybenzoic isomer II, and gallic acids. For the hydroxycinnamic acids, coumaric acid isomer II was the main moiety; ferulic acid and its hexoside, as well as the feruoylquinic acid isomer I, were also present in an abundant proportion. Aborus and collaborators [16] also identified these phenolic compounds in sprouted oats of the Golozrni and Jadar varieties, with *p*–hydroxybenzoic and sinapic acids being their principal acids; however, in the Turquesa oat variety, we additionally identified hexoside compounds, isoflavones, and phytosterols. In the same report, myricetin and kaempferol were the principal flavonoids, while in our study, hexoside forms of both compounds were predominant in the Turquesa sprouted oat variety. Oksman-Caldentey and collaborators [50] reported an increase in sterols of up to 20% during the germination process; however, in this study, we observed a reduction of around 15% of these compounds. On the other hand, Pecio and collaborators [58] reported the identification of avenacoside A in different oat seeds; similarly, we identified this saponin in the Turquesa oat variety and in its SO flour.

We used the combination of AOM and DSS to induce colorectal cancer over a short-term period in mice in order to evaluate the chemopreventive effect of whole oat sprouts (SO) or their phenolic AVA extract (AVA). The AOM + DSS-treated groups predominantly developed tumors in the distal zone of the colon, which is consistent with previous reports [39,40,59,60]. In this regard, tumor (macroscopic quantitative evaluation) and adenocarcinoma incidences (histopathological classification) induced by AOM + DSS were significantly correlated ($r = 0.97$, $p < 0.05$), confirming that these protuberant-type lesions were capable of developing into adenocarcinomas [2,60]. The fact that SO- and AVA-treated groups developed mostly pedunculate polyps is of major relevance, since sessile polyps are considered to have greater malignancy because they have a broad implantation base, without stems, on the surface of the mucosa, so that the degeneration of the cells reaches the base earlier, whereas pedunculate polyps represent a lower malignancy because the damage to the cells takes time to reach the support base [41]. Regarding flat-type lesions, animals treated with SO and its phenolic AVA extract behaved as expected, since a plaque is an early lesion of colonic mucosa that is not considered malignant [2,30,59]. Similarly, the medium inflammation grade and the low plaque incidence in the Normal group is due to the susceptibility of the strain to spontaneously developing these types of lesions [30].

Although the animals in the AOM + DSS-treated groups did not show any signs of colitis at the end of the experimental period, we still observed inflammation grades + and ++ in the colons of animals from AVA- and SO-treated groups, which were lesser grades in comparison with that of the AOM + DSS control group (+++). According to the literature [11–13], in our study, we observed a mild anti-inflammatory effect of the SO and AVA treatments. It has been reported that DSS administration in the drinking water triggers a state of chronic intestinal inflammation by binding to medium-chain-length fatty acids present in the mouse colon, inducing disruption of the colonic epithelial barrier [61]. In addition, DSS causes bloody diarrhea, ulcerations, and heavy infiltration of inflammatory cells into the mucosa [62], suggesting that inflammation is involved in the tumor-promotion activity of DSS. Similar results have been observed in our study and by others in the AOM/DSS-induced CRC mouse model [38–40,63,64].

Studies regarding the chemopreventive effect of oat sprouts and whole oats are limited in comparison with those discussing the unique polyphenolic alkaloids—AVAs—exclusively extracted from oats, among other compounds such as steroidal saponins, β-glucan, and flavonoids [5,9–11,13,15,44,45,52,65–68], mainly through mechanisms related to their antioxidant, anti-inflammatory, immunomodulatory, antiproliferative, proapoptotic, cancer cell growth, and senescence control activities. These studies suggest that the notion of effective antitumor activity arising from whole oats may be due to the synergistic effects of multiple compounds rather than any one nutrient or compound alone. In this regard, Wang and collaborators [69], by using the 1,2-dimethyl hydrazine (DMH)/DSS mouse colon cancer model, evaluated the preventive effect of whole-oat-containing diets. Their results indicated that low-, middle-, and high-dose whole oat diets (75, 150, and 225 g/kg of experimental diets, respectively) significantly reduced the number of aberrant crypt foci (ACF) and colon tumor incidence by 60%, 100%, and 100%, respectively, in comparison with DMH/DSS-induced ICR mice (with 60% tumor incidence), and significantly suppressed colon tumor growth in vivo. According to their dosage, the low-, middle-, and high-dose whole oat diets corresponded to 60, 120, and 180 mg of whole oats per mouse per day. Furthermore, here, we report the anticarcinogenic activity of sprouted oat (SO, 30 mg/day) and its phenolic AVA extract (AVA, 0.084 mg GAE/day) in the AOM/DSS-induced mouse colorectal carcinogenesis model by reducing the adenocarcinoma incidence by 52.5% and 21.3%, respectively. Overall, the results obtained in our study suggest that, despite the fact that inflammation was still presented in colonic samples of treated groups, compounds present in SO and its phenolic AVA extract could be blocking, inhibiting, or delaying the carcinogenesis process, particularly during the stage from promotion to transformation, which is where early lesions (plaques) evolve into neoplastic lesions that can become malignant [39,40,63]. Therefore, SO and AVA treatments were effective in reducing tumor and adenocarcinoma burdens, the compositions of resident inflammatory cells, and other inflammatory scores in the colons of AOM + DSS-treated mice, suggesting that oat products still retain their biological properties even after germination processing.

The absorption, bioavailability, and metabolism of several phytochemicals is a crucial factor in determining their biological activity against colon cancer. Fernaández-Ochoa and collaborators [23] examined at the in situ level the absorption, bioavailability, and metabolism of the phenolic compounds present in a rosemary leaf extract with proven antiproliferative and cytotoxic properties on colon cancer cells, and identified the main flavonoids, diterpenes, and triterpenes of the rosemary extract in gastrointestinal liquid and plasma samples together with metabolites from reactions of carnosic acid, carnosol, and rosmanol. In this body of evidence, it is hypothesized that the metabolites of flavonoids are primarily responsible for the observed anti-cancer effects owing to the unstable nature of the parent compounds at neutral or alkaline pH (pH > 8, as in the intestine) and their degradation by colonic microflora. Sankaranarayanan and collaborators [70] demonstrated the ability of the A-ring flavonoid metabolite, 2,4,6-trihydroxybenzoic acid (2,4,6-THBA), to inhibit Cyclin-Dependent Kinase (CDK) activity and cancer cell proliferation in colon cancer cell lines. Similarly, Wang and collaborators [20] investigated the biotransformation of AVA-C (2c) by mice and the human microbiota and found that AVA-2c and its major metabolite dihydroavenanthramide-C (M6) are bioactive compounds against human HCT-116 colon cancer cells through mechanisms related to apoptosis. On the other hand, the main mechanism involved in colon cancer prevention by oat β-glucan is the modulation of colon microbiota, which reduces the conversion of primary bile acids to secondary bile acids that are known to be tumorigenic. In addition, oat β-glucan administration increased the population of *Lactobacillus* and *Bifidobacterium*, but decreased the number of *Enterobacteriaceae* and induced a significant decrease in β-GA in rats. Moreover, oat β-glucan promotes the synthesis of short-chain fatty acids (SCFA), which are well-known anticarcinogenic compounds by colonic anaerobic bacteria and facilitate tumor cell apoptosis [9,10].

Colonic pH and β-GA activity are other physiological parameters involved in colorectal carcinogenesis. The increase in the pH of the colon and the activity of the β-GA enzyme is related to the specific colon carcinogen DMH and its metabolite AOM, as well as in the presence of the DSS promoter [39,60,63,71]. Therefore, inhibition of β-GA in cecal, colonic, and fecal samples of both SO- and AVA-treated groups is of major relevance, since β-glucuronidase is an enzyme also present in the human colonic microbiota which has the ability to hydrolyze many glucuronide conjugates and, as a consequence, can release active carcinogenic metabolites into the intestinal lumen [60]. In this regard, our results showed a positive correlation between β-GA cecal activity and the mean number of tumors and adenocarcinomas (r = 0.98 and r = 095, respectively, $p < 0.05$), which further supports the evidence that the activity of β-GA is involved in the development of colon cancer [72]; thus, modulation of β-GA through dietary treatment is confirmed as an efficient strategy for the prevention of this pathology.

It has been reported that increases in oxidative stress and/or decreases in antioxidant capacity are involved in the development of noncommunicable diseases such as cancer [42,73,74]. As a cellular defense mechanism, the Keap1 (Kelch-like ECH-associated protein 1)-Nrf2 (NF-E2-related factor 2) system was identified to respond to redox-disrupting stimuli, which directly modify Keap1, leading to inactivation of the Keap1 function, stabilization, and nuclear translocation of Nrf2, as well as induction of cytoprotective genes, such as those for GSH synthesis, oxidative stress elimination, detoxification, drug excretion, and anti-inflammatory response, among other functions [74]. In this regard, mounting evidence suggests that the increase of the antioxidant defense system, both by enzymatic and non-enzymatic means, has been reported in cancer tissues from patients and tissue samples from in vivo models of CCR, thus indicating an augmented defense against oxidative and inflammatory damage in cancer [73,75–77]. In our study, similar results were observed for serum GSH levels and GST and NQO1 enzymatic activities in the colons and livers of AOM + DSS-treated mice. In fact, AOM is a procarcinogen that undergoes oxidative metabolism in the liver, generating the production of active carcinogenic electrophiles (diazonium ion) that are released into the circulation and that eventually lead to peroxidation of plasma lipids and red blood cells (erythrocytes) [78]. In this regard, GSH is the main cellular defense against oxidative stress, and its antioxidant function is based on its ability to eliminate free radicals, reduce peroxides, and participate as a co-substrate in the activity of GSH-dependent enzymes, such as GST and glutathione peroxidase, among others [79]. As reported by Matić and collaborators [80], a lower serum concentration of GSH in AVA- and SO-treated animals, as compared to those of the AOM + DSS-treated group, may be due to the increased turnover of GSH in order to prevent oxidative damage, suggesting that GSH might have been used as an antioxidant to eliminate free radical and metabolite products of AOM/DSS that are conjugated with GSH before excretion to counteract lipid peroxidation and normalize oxidative stress in the bloodstream. Furthermore, in this study, a positive correlation was found between the content of GSH in erythrocytes and the means of polyps and adenocarcinomas ($r = 0.99$ and $r = 0.97$, $p < 0.05$, respectively), which indicates that endogenous antioxidant defense mechanisms are closely related to the incidence of early lesions and the development of adenocarcinomas in colon cancer.

Similarly, AVA and SO treatments achieved normalization of GST and NQO1 activities in the colon, suggesting that both treatments could neutralize the effects of oxidative stress at the colon level, initially generated by the oxidative metabolism of AOM and inflammation induced with DSS and, subsequently, by persistent inflammation in this disease model [75,76,81,82], resulting in a reduced carcinogenic impact. In the liver, AVA treatment normalized GST and NQO1 activities, while SO induced both activities. These results suggest that SO is more efficient in activating the Keap1-Nrf2 signaling pathway compared to treatment with AVA, which confirms that oat phenolic compounds together with β-glucans may be acting synergistically, thus offering greater protection for cancer prevention and treatment [65,83]. Tissue differences in GST activity might be attributed to cis elements in the promoter regions of these genes, known as the antioxidant/electrophile response elements, and the presence of the various transcription factors (members of the mammalian cap'n'collar (CNC) transcription factor family that possess a well-conserved basic region-leucine zipper (bZIP) motif) that heterodimerize with Nrf2, thus resulting in higher basal and inducible activities of GST in the liver as those compared to other tissues [74,84].

As mentioned above, the metabolism of flavonoids by gut microbiota is a crucial factor in determining their biological activity against colon cancer. Quercetin glycosides can be metabolized by intestinal bacteria into ring-fission products; the 3,4-dihydroxyphenylacetic acid (DOPAC) has recently been identified as the most active phenolic acid derived from quercetin glycosides, in terms of free-radical scavenging and induction of drug metabolizing enzymes. DOPAC simultaneously stimulates nuclear translocation of Nrf2 and aryl hydrocarbon receptor (AhR), both of which are responsible for the expression of phase I and II drug-metabolizing enzymes, such as GST and NQO1 [85]. Similarly, the oat-bran-derived phenolics ferulic and caffeic acids activate the Nrf2 signaling pathway [86,87].

Overall, the sprouting of seeds promotes degradation of macronutrients and antinutritional compounds and the biosynthesis of secondary metabolites with potential health benefits. These changes impact the nutritional value and health-promoting potential of edible seeds, such as those observed for germinated seeds and their derivatives of brown rice (*Oryza sativa*), rough rice, barley (*Hordeum vulgare*) foodstuff, and soybeans (*Glycine max* L.), from which in vivo and in vitro activities have been evaluated in order to determine their roles in CRC chemoprevention [88–93]. Here, we also provide evidence related to the chemopreventive effect of oat sprouts in AOM/DSS-induced mouse CRC, highlighting germination as a promising affordable food strategy for improving the potential health benefits of grains.

## 5. Conclusions

This work delivers information regarding the identification of the chemical families in Turquesa oat seeds previously reported in other oat varieties, their enhanced abundance differences compared to those of sprouted oats, and their association with the chemoprotective effect observed in this study. In addition, we identified chemical families not previously reported in other oat varieties that might contribute to the anticancer effect of germinated oats. More importantly, we here provide experimental evidence for a novel biological application of *Avena sativa*—Turquesa variety—sprouts in preventing colon cancer, and we identify the contribution of the sprout's phenolic AVA extract to achieving this effect, which is a result of its antioxidant activity, reduction of inflammatory status, and improvement of colonic physiological parameters. The major relevance of our study was the superior chemopreventive effect of the sprouted oat, probably due to the synergistic effects of multiple compounds. This further supports the potential use of oat sprouts as a functional food for colon cancer prevention.

Future investigations will now focus on the identification and quantitation of the bioactive compounds at different germination conditions to enhance the phytochemical profiles and the biological activities of oat sprouts. In addition, the examination of the absorption, bioavailability, and metabolism at several levels of the chemical families present in the sprouted oat will be of major relevance in order to clarify the absorption and metabolism of sprouted oat bioactive compounds, which in turn would contribute to a fuller understanding of the mechanisms of action of these compounds against colorectal cancer.

**Supplementary Materials:** Supplementary materials can be found at http://www.mdpi.com/2304-8158/9/2/169/s1. Table S1: Flavonoid and non-flavonoid profile of oat seed and SO obtained by the UPLC-QTOF analysis.

**Author Contributions:** Conceptualization, M.R.-G. and R.R.-C.; Formal analysis, M.D.-L. and G.R.-P.; Funding acquisition, M.R.-G. and R.R.-C.; Investigation, M.D.-L., I.F.P.-R., and E.A.d.l.R.; Methodology, M.D.-L., G.R.-P., I.F.P.-R., N.E.R.-G., and E.A.d.l.R.; Project administration, M.D.-L. and M.R.-G.; Resources, M.R.-G., N.E.R.-G., and R.R.-C.; Supervision, R.R.-C. and M.R.-G.; Validation, I.F.P.-R. and N.E.R.-G.; Visualization, M.D.-L.; Writing—original draft, M.D.-L.; Writing—review & editing, M.R.-G. All authors read, provided comments, and approved the manuscript. All authors have read and agreed to the published version of the manuscript.

**Funding:** The research was founded by Fondo de Proyectos Especiales de Rectoría (FOPER-UAQ-2018). Project code: FOPER-2019-01031.

**Acknowledgments:** The authors are grateful to The National Council of Science and Technology, Mexico (CONACYT) for contributing a scholarship grant for M.D.-L.; Instituto Nacional de Investigaciones Forestales, Agrícolas y Pecuarias (INIFAP) Campo Experimental Bajío, Celaya, GT, MEX and Acosta, J for donating oat (*Avena sativa* L.) seeds, "Turquesa" variety. We appreciate the technical assistance of the Microscopy Unit of the Institute of Neurobiology UNAM–Juriquilla (QT, MEX); Hérnandez, E.N.; Gallejos, M.A and the animal care by MVZ. Garcia, J.M.

**Conflicts of Interest:** The authors declare no conflict of interest.

## Abbreviations

| | |
|---|---|
| ACF | Aberrant crypt foci |
| AOM | Azoxymethane |
| AVA | Avenanthramides |
| β–GA | β–glucuronidase activity |
| CRC | Colorectal cancer |
| DMH | 1,2–Dimethylhydrazine |
| DSS | Dextran sulfate sodium |
| GSH | Glutathione reduced |
| GST | Glutathione *S*–transferase |
| H&E | Hematoxylin and eosin |
| NQO1 | NADPH:Quinone oxidoreductase 1 |
| RH | Relative humidity |
| SO | Sprouted oat |
| TPC | Total phenolic compounds |

## References

1. Arnold, M.; Sierra, M.S.; Laversanne, M.; Soerjomataram, I.; Jemal, A.; Bray, F. Global patterns and trends in colorectal cancer incidence and mortality. *Gut* **2017**, *66*, 683–691. [CrossRef]
2. Dulal, S.; Keku, T.O. Gut microbiome and colorectal adenomas. *Cancer J.* **2014**, *20*, 225–231. [CrossRef]
3. Marshall, J.R. Prevention of colorectal cancer: Diet, chemoprevention, and lifestyle. *Gastroenterol. Clin. N. Am.* **2008**, *37*, 73–82. [CrossRef]
4. Cappellani, A.; Zanghì, A.; Di Vita, M.; Cavallaro, A.; Piccolo, G.; Veroux, P.; Lo Menzo, E.; Cavallaro, V.; de Paoli, P.; Veroux, M.; et al. Strong correlation between diet and development of colorectal cancer. *Front. Biosci.* **2013**, *18*, 190–198. [CrossRef]
5. Kasdagly, M.; Radhakrishnan, S.; Reddivari, L.; Veeramachaneni, D.R.; Vanamala, J. Colon carcinogenesis: Influence of Western diet-induced obesity and targeting stem cells using dietary bioactive compounds. *Nutrition* **2014**, *30*, 1242–1256. [CrossRef]
6. Murphy, J.K.; Marques-Lopes, I.; Sánchez-Tainta, A. Cereals and Legumes. In *The Prevention of Cardiovascular Disease through the Mediterranean Diet*, 1st ed.; Sánchez-Villegas, A., Sanchez-Taínta, A., Eds.; Academic Press: Waltham, MA, USA, 2018; pp. 111–132. [CrossRef]
7. Arendt, E.K.; Zannini, E. Oats. In *Woodhead Publishing Series in Food Science, Technology and Nutrition, Cereal Grains for the Food and Beverage Industries*, 1st ed.; Arendt, E.K., Zannini, E., Eds.; Woodhead Publishing: Cambridge, UK, 2013; pp. 243–283. [CrossRef]
8. Villaseñor, H.E.; Espitia, R.E.; Huerta, E.J.; Osorio, A.L.; López, H.J. Turquesa, nueva variedad de avena para la producción de grano y forraje en México. *Agric. Técnica México* **2009**, *35*, 487–492.
9. Martínez-Villaluenga, C.; Peñas, E. Health benefits of oat: Current evidence and molecular mechanisms. *Curr. Opin. Food Sci.* **2017**, *14*, 26–31. [CrossRef]
10. Shen, R.L.; Wang, Z.; Dong, J.L.; Xiang, Q.S.; Liu, Y.Q. Effects of oat soluble and insoluble β-glucan on 1,2-dimethylhydrazine-induced early colon carcinogenesis in mice. *Food Agric. Immunol.* **2016**, *27*, 657–666. [CrossRef]
11. Liu, B.; Lin, Q.; Yang, T.; Zeng, L.; Shi, L.; Chen, Y.; Luo, F. Oat β-glucan ameliorates dextran sulfate sodium (DSS)-induced ulcerative colitis in mice. *Food Funct.* **2015**, *6*, 3454–3463. [CrossRef]
12. Shahidi, F.; de Camargo, A.C. Tocopherols and tocotrienols in common and emerging dietary sources: Occurrence, applications, and health benefits. *Int. J. Mol. Sci.* **2016**, *17*, 1745. [CrossRef]
13. Yang, J.; Ou, B.; Wise, M.L.; Chu, Y. *In vitro* total antioxidant capacity and anti-inflammatory activity of three common oat-derived avenanthramides. *Food Chem.* **2014**, *160*, 338–345. [CrossRef] [PubMed]
14. Chen, C.; Wang, L.; Wang, R.; Luo, X.; Li, Y.; Li, J.; Li, Y.; Chen, Z. Phenolic contents, cellular antioxidant activity and antiproliferative capacity of different varieties of oats. *Food Chem.* **2018**, *239*, 260–267. [CrossRef] [PubMed]
15. Fu, R.; Yang, P.; Sajid, A.; Li, Z. Avenanthramide A Induces Cellular Senescence via miR-129-3p/Pirh2/p53 signaling pathway to suppress colon cancer growth. *J. Agric. Food Chem.* **2019**, *67*, 4808–4816. [CrossRef]
16. Aborus, N.E.; Šaponjac, V.T.; Ćanadanović-Brunet, J.; Ćetković, G.; Hidalgo, A.; Vulić, J.; Šeregelj, V. Sprouted and freeze-dried wheat and oat seeds–phytochemical profile and *in vitro* biological activities. *Chem. Biodivers.* **2018**, *15*, e1800119. [CrossRef]
17. Kaukovirta-Norja, A.; Wilhelmson, A.; Poutanen, K. Germination: A means to improve the functionality of oat. *Agric. Food Sci.* **2004**, *13*, 100–112. [CrossRef]
18. Dhakal, H.; Yang, E.J.; Lee, S.; Kim, M.J.; Baek, M.C.; Lee, B.; Park, P.H.; Kwon, T.K.; Khang, D.; Song, K.S.; et al. Avenanthramide C from germinated oats exhibits anti-allergic inflammatory effects in mast cells. *Sci. Rep.* **2019**, *9*, 6884. [CrossRef]
19. Koenig, R.T.; Dickman, J.R.; Wise, M.L.; Ji, L.L. Avenanthramides are bioavailable and accumulate in hepatic, cardiac, and skeletal muscle tissue following oral gavage in rats. *J. Agric. Food Chem.* **2011**, *59*, 6438–6443. [CrossRef]
20. Wang, P.; Chen, H.; Zhu, Y.; McBride, J.; Fu, J.; Sang, S. Oat avenanthramide-C (2c) is biotransformed by mice and the human microbiota into bioactive metabolites. *J. Nutr.* **2014**, *145*, 239–245. [CrossRef]
21. Chen, C.Y.; Milbury, P.E.; Kwak, H.K.; Collins, F.W.; Samuel, P.; Blumberg, J.B. Avenanthramides and phenolic acids from oats are bioavailable and act synergistically with vitamin C to enhance hamster and human LDL resistance to oxidation. *J. Nutr.* **2004**, *134*, 1459–1466. [CrossRef]

22. Chen, C.Y.; Milbury, P.E.; Collins, F.W.; Blumberg, J.B. Avenanthramides are bioavailable and have antioxidant activity in humans after acute consumption of an enriched mixture from oats. *J. Nutr.* **2007**, *137*, 1375–1382. [CrossRef]
23. Fernaández-Ochoa, A.; Borraás-Linares, I.; Peárez-Saánchez, A.; Barrajoán-Catalaán, E.; Gonzaález-Aálvarez, I.; Arraáez-Romaán, D.; Micol, V.; Segura-Carretero, A. Phenolic compounds in rosemary as potential source of bioactive compounds against colorectal cancer: *In situ* absorption and metabolism study. *J. Funct. Foods* **2017**, *33*, 202–210. [CrossRef]
24. Horwitz, W. AOAC International. In *Official Methods of Analysis of AOAC International*, 17th ed.; Horwitz, W., AOAC International, Eds.; AOAC International Seventeen: Gaithersburg, MD, USA, 2002.
25. McCleary, B.V.; Codd, R. Measurement of (1→ 3), (1→ 4)-β-D-glucan in barley and oats: A streamlined enzymic procedure. *J. Sci. Food Agric.* **1991**, *55*, 303–312. [CrossRef]
26. Latta, M.; Eskin, M. A simple and rapid colorimetric method for phytate determination. *J. Agric. Food Chem.* **1980**, *28*, 1313–1315. [CrossRef]
27. Singleton, V.L.; Rossi, J.A. Colorimetry of total phenolics with phosphomolybdic-phosphotungstic acid reagents. *Am. J. Enol. Vitic.* **1965**, *16*, 144–158.
28. Rodríguez-González, S.; Pérez-Ramírez, I.F.; Castaño-Tostado, E.; Amaya-Llano, S.; Rodríguez-García, M.E.; Reynoso-Camacho, R. Improvement of physico-chemical properties and phenolic compounds bioavailability by concentrating dietary fiber of peach (*Prunus persica*) juice by-product. *J. Sci. Food Agric.* **2018**, *98*, 3109–3118. [CrossRef]
29. Jenab, M.; Thompson, L.U. The influence of flaxseed and lignans on colon carcinogenesis and β-glucuronidase activity. *Carcinogenesis* **1996**, *17*, 1343–1348. [CrossRef]
30. Astler, V.B.; Coller, F.A. The prognostic significance of direct extension of carcinoma of the colon and rectum. *Ann. Surg.* **1954**, *139*, 846–852. [CrossRef]
31. Riddell, R.H.; Goldman, H.; Ransohoff, D.F.; Appelman, H.D.; Fenoglio, C.M.; Haggitt, R.C.; Ahren, C.; Correa, P.; Hamilton, S.R.; Morson, B.C. Dysplasia in inflammatory bowel disease: Standardized classification with provisional clinical applications. *Hum. Pathol.* **1983**, *14*, 931–968. [CrossRef]
32. Klintrup, K.; Mäkinen, J.M.; Kauppila, S.; Väre, P.O.; Melkko, J.; Tuominen, H.; Tuppurainen, K.; Mäkela, J.; Karttunen, T.J.; Mäkinen, M.J. Inflammation and prognosis in colorectal cancer. *Eur. J. Cancer* **2005**, *41*, 2645–2654. [CrossRef]
33. Fleming, M.; Ravula, S.; Tatishchev, S.F.; Wang, H.L. Colorectal carcinoma: Pathologic aspects. *J. Gastrointest. Oncol.* **2012**, *3*, 153–173. [CrossRef]
34. Marchal-Bressenot, A.; Riddell, R.H.; Boulagnon-Rombi, C.; Reinisch, W.; Danese, S.; Schreiber, S.; Peyrin-Biroulet, L. Review article: The histological assessment of disease activity in ulcerative colitis. *Aliment. Pharmacol. Ther.* **2015**, *42*, 957–967. [CrossRef]
35. Ellman, G.L. Tissue sulfhydryl groups. *Arch. Biochem. Biophys.* **1959**, *82*, 70–77. [CrossRef]
36. Habig, W.H.; Pabst, M.J.; Jakoby, W.B. Glutathione S-transferases. The first enzymatic step in mercapturic acid formation. *J. Biol. Chem.* **1974**, *249*, 7130–7139.
37. Prochaska, H.J.; Santamaria, A.B.; Talalay, P. Rapid detection of inducers of enzymes that protect against carcinogens. *Proc. Natl. Acad. Sci. USA* **1992**, *89*, 2394–2398. [CrossRef]
38. Valadez-Bustos, N.; Escamilla-Silva, E.M.; García-Vázquez, F.J.; Gallegos-Corona, M.A.; Amaya-Llano, S.L.; Ramos-Gómez, M. Oral administration of microencapsulated *B. longum* BAA-999 and lycopene modulates IGF-1/IGF-1R/IGFBP3 protein expressions in a colorectal murine model. *Int. J. Mol. Sci.* **2019**, *20*, 4275. [CrossRef]
39. Tanaka, T.; Kohno, H.; Suzuki, R.; Yamada, Y.; Sugie, S.; Mori, H. A novel inflammation-related mouse colon carcinogenesis model induced by azoxymethane and dextran sodium sulfate. *Cancer Sci.* **2003**, *94*, 965–973. [CrossRef]
40. Tanaka, T. Development of an inflammation-associated colorectal cancer model and its application for research on carcinogenesis and chemoprevention. *Int. J. Inflam.* **2012**, *2012*, 658786. [CrossRef]
41. Neergheen, V.S.; Bahorun, T.; Taylor, E.W.; Jen, L.S.; Aruoma, O.I. Targeting specific cell signaling transduction pathways by dietary and medicinal phytochemicals in cancer chemoprevention. *Toxicology* **2010**, *278*, 229–241. [CrossRef]
42. Aune, D. Plant Foods, antioxidant biomarkers, and the risk of cardiovascular disease, cancer, and mortality: A review of the evidence. *Adv. Nutr.* **2019**, *10*, S404–S421. [CrossRef]

43. Schwingshackl, L.; Morze, J.; Hoffmann, G. Mediterranean diet and health status: Active ingredients and pharmacological mechanisms. *Br. J. Pharmacol.* **2019**, 1–17. [CrossRef]
44. de Bruijn, W.J.; van Dinteren, S.; Gruppen, H.; Vincken, J.P. Mass spectrometric characterisation of avenanthramides and enhancing their production by germination of oat (*Avena sativa*). *Food Chem.* **2019**, *277*, 682–690. [CrossRef]
45. Lee, J.H.; Lee, B.K.; Park, H.H.; Lee, B.W.; Woo, K.S.; Kim, H.J.; Han, S.I.; Lee, Y.Y. Oat germination and ultrafiltration process improves the polyphenol and avenanthramide contents with protective effect in oxidative-damaged HepG2 cells. *J. Food Biochem.* **2019**, *43*, e12799. [CrossRef]
46. Leonova, S.; Grimberg, A.; Marttila, S.; Stymne, S.; Carlsson, A.S. Mobilization of lipid reserves during germination of oat (*Avena sativa* L.), a cereal rich in endosperm oil. *J. Exp. Bot.* **2010**, *61*, 3089–3099. [CrossRef]
47. Lehtinen, P.; Kaukovirta-Norja, A. Oats: Chemistry and Technology. In *Oat Lipids, Enzymes, and Quality*; Webster, F.H., Wood, P.J., Eds.; AACC International: Saint Paul, MN, USA, 2011; pp. 143–156. [CrossRef]
48. Menon, R.; Gonzalez, T.; Ferruzzi, M.; Jackson, E.; Winderl, D.; Watson, J. Oats-from farm to fork. *Adv. Food. Nutr. Res.* **2016**, *77*, 1–55. [CrossRef]
49. Welch, R.W. Nutrient composition and nutritional quality of oats and comparisons with other cereals. In *Oats: Chemistry and Technology*; Webster, F.H., Wood, P.J., Eds.; AACC International: Saint Paul, MN, USA, 2011; pp. 95–107.
50. Oksman-Caldentey, K.M.; Kaukovirta-Norja, A.; Heiniö, R.-L.; Kleemola, T.; Mikola, M.; Sontag-Strom, T.; Lehtinen, P.; Pihlava, J.M.; Poutanen, P. Kauran biotekninen prosessointi uusiksi elintarvikkeiksi. In *Kau-Ran Biotekninen Prosessointi Uusiksi Elintarvikkeiksi (Biotechnical Processing of Oat for Novel Food Ingredients)*; Salovaara, H., Sontag-Strom, T., Eds.; EKT-Sarja 1221: Elintarviketeknologian Laitos, Finland, 2001; pp. 85–108.
51. Silva, E.O.; Bracarense, A.P. Phytic acid: From antinutritional to multiple protection factor of organic systems. *J. Food Sci.* **2016**, *81*, R1357–R1362. [CrossRef]
52. Boz, H. Phenolic amides (avenanthramides) in oats—A review. *Czech J. Food Sci.* **2015**, *33*, 399–404. [CrossRef]
53. Collins, F.W. Oat phenolics: Avenanthramides, novel substituted N-cinnamoylanthranilate alkaloids from oat groats and hulls. *J. Agric. Food Chem.* **1989**, *37*, 60–66. [CrossRef]
54. Emmons, C.L.; Peterson, D.M. Antioxidant activity and phenolic contents of oat groats and hulls. *Cereal Chem.* **1999**, *76*, 902–906. [CrossRef]
55. Emmons, C.L.; Peterson, D.M. Antioxidant activity and phenolic content of oat as affected by cultivar and location. *Crop Sci.* **2001**, *41*, 1676–1681. [CrossRef]
56. Dimberg, L.H.; Gissén, C.; Nilsson, J. Phenolic compounds in oat grains (*Avena sativa* L.) grown in conventional and organic systems. *AMBIO J. Hum. Environ.* **2005**, *34*, 331–337. [CrossRef]
57. McMurrough, I.; Baert, T. Identification of proanthocyanidins in beer and their direct measurement with a dual electrode electrochemical detector. *J. Inst. Brew.* **1994**, *100*, 409–416. [CrossRef]
58. Pecio, Ł.; Wawrzyniak-Szołkowska, A.; Oleszek, W.; Stochmal, A. Rapid analysis of avenacosides in grain and husks of oats by UPLC-TQ-MS. *Food Chem.* **2013**, *141*, 2300–2304. [CrossRef] [PubMed]
59. Tanaka, T. Colorectal carcinogenesis: Review of human and experimental animal studies. *J. Carcinog.* **2009**, *8*, 5. [CrossRef]
60. Perše, M.; Cerar, A. Morphological and molecular alterations in 1,2 dimethylhydrazine and azoxymethane induced colon carcinogenesis in rats. *BioMed Res. Int.* **2011**, *2011*, 473964. [CrossRef]
61. Gkouskou, K.K.; Deligianni, C.; Tsatsanis, C.; Eliopoulos, A.G. The gut microbiota in mouse models of inflammatory bowel disease. *Front. Cell. Infect. Microbiol.* **2014**, *4*, 28. [CrossRef]
62. Grimm, V.; Radulovic, K.; Riedel, C.U. Colonization of C57BL/6 Mice by a potential probiotic *Bifidobacterium bifidum* strain under germ-free and specific pathogen-free conditions and during experimental colitis. *PLoS ONE* **2015**, *10*, e0139935. [CrossRef]
63. de Robertis, M.; Massi, E.; Poeta, M.L.; Carotti, S.; Morini, S.; Cecchetelli, L.; Signori, E.; Fazio, V.M. The AOM/DSS murine model for the study of colon carcinogenesis: From pathways to diagnosis and therapy studies. *J. Carcinog.* **2011**, *10*, 9. [CrossRef]
64. Doulberis, M.; Angelopoulou, K.; Kaldrymidou, E.; Tsingotjidou, A.; Abas, Z.; Erdman, S.E.; Poutahidis, T. Cholera-toxin suppresses carcinogenesis in a mouse model of inflammation-driven sporadic colon cancer. *Carcinogenesis* **2015**, *36*, 280–290. [CrossRef]

65. Suchecka, D.; Harasym, J.; Wilczak, J.; Gromadzka-Ostrowska, J. Hepato- and gastro-protective activity of purified oat 1-3, 1-4-β-d-glucans of different molecular weight. *Int. J. Biol. Macromol.* **2016**, *91*, 1177–1185. [CrossRef]
66. Yang, J.; Wang, P.; Wu, W.; Zhao, Y.; Idehen, E.; Sang, S. Steroidal Saponins in oat bran. *J. Agric. Food Chem.* **2016**, *64*, 1549–1556. [CrossRef]
67. Hazafa, A.; Rehman, K.U.; Jahan, N.; Jabeen, Z. The role of polyphenol (Flavonoids) compounds in the treatment of cancer cells. *Nutr. Cancer* **2019**, *9*, 1–12. [CrossRef]
68. Turrini, E.; Maffei, F.; Milelli, A.; Calcabrini, C.; Fimognari, C. Overview of the anticancer profile of avenanthramides from oat. *Int. J. Mol. Sci.* **2019**, *20*, 4536. [CrossRef]
69. Wang, H.C.; Hung, C.H.; Hsu, J.D.; Yang, M.Y.; Wang, S.J.; Wang, C.J. Inhibitory effect of whole oat on aberrant crypt foci formation and colon tumor growth in ICR and BALB/c mice. *J. Cereal Sci.* **2011**, *53*, 73–77. [CrossRef]
70. Sankaranarayanan, R.; Valiveti, C.K.; Kumar, D.R.; van slambrouck, S.; Kesharwani, S.S.; Seefeldt, T.; Scaria, J.; Tummala, H.; Bhat, G.J. The flavonoid metabolite 2,4,6-trihydroxybenzoic acid is a CDK inhibitor and an anti-proliferative agent: A potential role in cancer prevention. *Cancers* **2019**, *11*, 427. [CrossRef]
71. Cuellar-Nuñez, M.L.; Luzardo-Ocampo, I.; Campos-Vega, R.; Gallegos-Corona, M.A.; de Mejía, E.G.; Loarca-Piña, G. Physicochemical and nutraceutical properties of moringa (Moringa oleifera) leaves and their effects in an in vivo AOM/DSS-induced colorectal carcinogenesis model. *Food Res. Int.* **2018**, *105*, 159–168. [CrossRef]
72. Cheng, C.M.; Chen, F.M.; Lu, Y.L.; Tzou, S.C.; Wang, J.Y.; Kao, C.H.; Liao, K.W.; Cheng, T.C.; Chuang, C.H.; Chen, B.M.; et al. Expression of β-glucuronidase on the surface of bacteria enhances activation of glucuronide prodrugs. *Cancer Gene Ther.* **2013**, *20*, 276–281. [CrossRef]
73. Mandal, P. Potential biomarkers associated with oxidative stress for risk assessment of colorectal cancer. *Naunyn-Schmiedeberg's Arch. Pharmacol.* **2017**, *390*, 557–565. [CrossRef]
74. Yamamoto, M.; Kensler, T.W.; Motohashi, H. The KEAP1-NRF2 system: A thiol-based sensor-effector apparatus for maintaining redox homeostasis. *Physiol. Rev.* **2018**, *98*, 1169–1203. [CrossRef]
75. Pool-Zobel, B.; Veeriah, S.; Böhmer, F.D. Modulation of xenobiotic metabolising enzymes by anticarcinogens—Focus on glutathione S-transferases and their role as targets of dietary chemoprevention in colorectal carcinogenesis. *Mutat. Res.-Fundam. Mol. Mech. Mutagenesis* **2005**, *591*, 74–92. [CrossRef]
76. Strzelczyk, J.K.; Wielkoszyński, T.; Krakowczyk, Ł.; Adamek, B.; Zalewska-Ziob, M.; Gawron, K.; Kasperczyk, J.; Wiczkowski, A. The activity of antioxidant enzymes in colorectal adenocarcinoma and corresponding normal mucosa. *Acta Biochim. Pol.* **2012**, *59*, 549–556. [CrossRef]
77. Lippmann, D.; Lehmann, C.; Florian, S.; Barknowitz, G.; Haack, M.; Mewis, I.; Wiesner, M.; Schreiner, M.; Glatt, H.; Brigelius-Floh´e, R.; et al. Glucosinolates from pak choi and broccoli induce enzymes and inhibit inflammation and colon cancer differently. *Food Funct.* **2014**, *5*, 1073–1081. [CrossRef]
78. Bobek, P.; Galbavý, Š.; Mariassyova, M. The effect of red beet (*Beta vulgaris var. rubra*) fiber on alimentary hypercholesterolemia and chemically induced colon carcinogenesis in rats. *Nahrung* **2000**, *44*, 184–187. [CrossRef]
79. Aquilano, K.; Baldelli, S.; Ciriolo, M.R. Glutathione: New roles in redox signaling for an old antioxidant. *Front. Pharmacol.* **2014**, *5*, 196. [CrossRef]
80. Matić, M.M.; Paunović, M.G.; Milošević, M.D.; Ognjanović, B.I.; Saičić, Z.S. Hematoprotective effects and antioxidant properties of β-glucan and vitamin C against acetaminophen-induced toxicity: An experimental study in rats. *Drug Chem. Toxicol.* **2019**, 1–8. [CrossRef]
81. Manju, V.; Balasubramaniyan, V.; Nalini, N. Rat colonic lipid peroxidation and antioxidant status: The effects of dietary luteolin on 1, 2-dimethylhydrazine challenge. *Cell. Mol. Biol. Lett.* **2005**, *10*, 535–551.
82. Begleiter, A.; Sivananthan, K.; Lefas, G.M.; Maksymiuk, A.W.; Bird, R.P. Inhibition of colon carcinogenesis by post-initiation induction of NQO1 in Sprague-Dawley rats. *Oncol. Rep.* **2009**, *21*, 1559–1565. [CrossRef]
83. Wu, Z.; Ming, J.; Gao, R.; Wang, Y.; Liang, Q.; Yu, H.; Zhao, G. Characterization and antioxidant activity of the complex of tea polyphenols and oat β-glucan. *J. Agric. Food Chem.* **2011**, *59*, 10737–10746. [CrossRef]
84. Ramos-Gomez, M.; Kwak, M.K.; Dolan, P.M.; Itoh, K.; Yamamoto, M.; Talalay, P.; Kensler, T.W. Sensitivity to carcinogenesis is increased and chemoprotective efficacy of enzyme inducers is lost in nrf2 transcription factor-deficient mice. *Proc. Natl. Acad. Sci. USA* **2001**, *98*, 3410–3415. [CrossRef]

85. Murota, K.; Nakamura, Y.; Uehara, M. Flavonoid metabolism: The interaction of metabolites and gut microbiota. *Biosci. Biotechnol. Biochem.* **2018**, *82*, 600–610. [CrossRef]
86. Wang, Y.; Huo, Y.; Zhao, L.; Lu, F.; Wang, O.; Yang, X.; Ji, B.; Zhou, F. Cyanidin-3-glucoside and its phenolic acid metabolites attenuate visible light-induced retinal degeneration in vivo via activation of Nrf2/HO-1 pathway and NF-κB suppression. *Mol. Nutr. Food Res.* **2016**, *60*, 1564–1577. [CrossRef]
87. Yang, S.Y.; Pyo, M.C.; Nam, M.H.; Lee, K.W. ERK/Nrf2 pathway activation by caffeic acid in HepG2 cells alleviates its hepatocellular damage caused by t-butylhydroperoxide-induced oxidative stress. *BMC Complement. Altern. Med.* **2019**, *19*, 139. [CrossRef] [PubMed]
88. Latifah, S.Y.; Armania, N.; Tze, T.H.; Azhar, Y.; Nordiana, A.H.; Norazalina, S.; Hairuszah, I.; Saidi, M.; Maznah, I. Germinated brown rice (GBR) reduces the incidence of aberrant crypt foci with the involvement of beta-catenin and COX-2 in azoxymethane-induced colon cancer in rats. *Nutr. J.* **2010**, *26*, 9–16. [CrossRef]
89. Saki, E.; Saiful-Yazan, L.; Mohd-Ali, R.; Ahmad, Z. Chemopreventive effects of germinated rough rice crude extract in inhibiting azoxymethane-induced aberrant crypt foci formation in Sprague-Dawley rats. *BioMed Res. Int.* **2017**, *2017*, 9517287. [CrossRef]
90. Kanauchi, O.; Mitsuyama, K.; Andoh, A.; Iwanaga, T. Modulation of intestinal environment by prebiotic germinated barley foodstuff prevents chemo-induced colonic carcinogenesis in rats. *Oncol. Rep.* **2008**, *20*, 793–801. [CrossRef]
91. Komiyama, Y.; Mitsuyama, K.; Masuda, J.; Yamasaki, H.; Takedatsu, H.; Andoh, A.; Tsuruta, O.; Fukuda, M.; Kanauchi, O. Prebiotic treatment in experimental colitis reduces the risk of colitic cancer. *J. Gastroenterol. Hepatol.* **2011**, *26*, 1298–1308. [CrossRef]
92. Mollah, M.L.; Park, D.K.; Park, H.J. *Cordyceps militaris* grown on germinated soybean induces G2/M cell cycle arrest through downregulation of cyclin B1 and Cdc25c in human colon cancer HT-29 Cells. *Evid. Based Complement. Altern. Med.* **2012**, *2012*, 249217. [CrossRef]
93. González-Montoya, M.; Hernández-Ledesma, B.; Silván, J.M.; Mora-Escobedo, R.; Martínez-Villaluenga, C. Peptides derived from *in vitro* gastrointestinal digestion of germinated soybean proteins inhibit human colon cancer cells proliferation and inflammation. *Food Chem.* **2018**, *1*, 75–82. [CrossRef]

Article

# Biochemical Properties of Polyphenol Oxidases from Ready-to-Eat Lentil (*Lens culinaris* Medik.) Sprouts and Factors Affecting Their Activities: A Search for Potent Tools Limiting Enzymatic Browning

**Małgorzata Sikora, Michał Świeca *, Monika Franczyk, Anna Jakubczyk, Justyna Bochnak and Urszula Złotek**

Department of Biochemistry and Food Chemistry, University of Life Sciences, Skromna Str. 8, 20-704 Lublin, Poland; malgorzata.sikora@up.lublin.pl (M.S.); monika.franczyk@up.lublin.pl (M.F.); anna.jakubczyk@up.lublin.pl (A.J.); just.bochnak@gmail.com (J.B.); urszula.zlotek@up.lublin.pl (U.Z.)

* Correspondence: michal.swieca@up.lublin.pl; Tel.: +48-81-4623327; Fax: +48-81-4623324

Received: 12 March 2019; Accepted: 6 May 2019; Published: 7 May 2019

**Abstract:** Enzymatic browning of sprouts during storage is a serious problem negatively influencing their consumer quality. Identifying and understanding the mechanism of inhibition of polyphenol oxidases (PPOs) in lentil sprouts may offer inexpensive alternatives to prevent browning. This study focused on the biochemical characteristics of PPOs from stored lentil sprouts, providing data that may be directly implemented in improving the consumer quality of sprouts. The purification resulted in approximately 25-fold enrichment of two PPO isoenzymes (PPO I and PPO II). The optimum pH for total PPOs, as well as for PPO I and PPO II isoenzymes, was 4.5–5.5, 4.5–5.0, and 5.5, respectively. The optimal temperature for PPOs was 35 °C. Total PPOs and the PPO I and PPO II isoenzymes had the greatest affinity for catechol ($K_m$ = 1.32, 1.76, and 0.94 mM, respectively). Ascorbic acid was the most effective in the inhibition of dark color formation by total PPOs, and showed ca. 62%, 43%, and 24% inhibition at 20-, 2-, and 0.2-mM concentrations. Ascorbic acid, L-cysteine, and sodium metabisulfite (20 mM) significantly inhibited color development in the reactions catalyzed by both isoenzymes of PPO. $Ba^{2+}$, $Fe^{3+}$, and $Mn^{2+}$ (10 mM) completely inhibited PPO activity. This study of the effect of antibrowning compounds and cations on PPO activity provides data that can be used to protect lentil sprouts against enzymatic browning during storage and processing.

**Keywords:** biochemical characteristic; enzymatic browning; inhibitory profile; lentil; sprouts; polyphenol oxidase; purification

## 1. Introduction

Polyphenol oxidases (PPOs) (EC 1.14.18.1, EC 1.10.3.1, or EC 1.10.3.2) are widely distributed in the plant kingdom, and their level and activity are dependent on the age, species, variety, maturity, and stress status of plants [1,2]. In addition, they are located in certain organelles, such as chloroplast thylakoids, peroxisomes, and mitochondria [1]. According to substrate specificity, three main types of phenol oxidases are known: (I) Monophenol monooxygenase (also called tyrosinase, monophenol oxidase, or cresolase) catalyzes the hydroxylation of monophenol to ortho-diphenol and the oxidation of diphenol to ortho-quinone; (II) diphenol oxidase (also called catechol oxidase, polyphenol oxidase, or *o*-diphenolase) catalyzes the oxidation of ortho-phenol, but cannot catalyze the oxidation or monooxygenation of metaphenol and para-phenol; and (III) laccase catalyzes the oxidation of ortho-phenol and para-phenol, but cannot catalyze the oxidation of monophenol and metaphenol [3]. This classification, although commonly used, also has some inaccuracies, e.g., in the case of mung bean [4] or tobacco [5] laccases, which share many substrates with PPOs. Enzymatic browning of plant-derived foods (including sprouts) contributes to a decrease in the sensory properties and marketability of fruits and vegetables [6–9]. The formation of brown or black pigments is due to increased activity of PPOs, resulting in the polymerization of quinones [2,7]. Additionally, increased activity of PPOs can decrease the level of phenolic compounds, i.e., plant secondary metabolites with well-documented nutraceutical properties [10,11]. Due to these facts, the characterization of PPO activities or the removal of reactants such as oxygen and phenolic compounds, especially those concerning potential inhibitors, are of increasing interest in the food industry. As one of the antibrowning agents, ascorbic acid inactivates PPOs irreversibly in the absence of PPO substrates, probably through binding to its active site, preferentially in its oxy form. Additionally, it can reduce reaction products, limiting the formation of a dark color. Cysteine activity is usually attributed to various mechanisms, e.g., its nucleophilic reactivity toward quinones to give a colorless adduct or its ability to reduce *o*-quinones to their polyphenol precursors. Citric acid and ethylenediaminetetraacetic acid sodium salt (EDTA) chelate copper at an enzyme-active site [12,13].

Polyphenol oxidase has been widely studied in fruits, vegetables, and mushrooms, e.g., eggplant (*Solanum melongena*) [14], persimmon [15], broccoli (*Brassica oleracea* var. *botrytis italica*) [16], celery [17], and butter lettuce (*Lactuca sativa* var. *capitata* L.) [18]: However, there are very few data that have presented the characterization of PPOs from edible sprouts.

In this paper, we report the isolation, partial purification, and biochemical properties of two isoenzymes and total PPO activity in lentil sprouts (*Lens culinaris* Medik.). Special attention is placed on the factors affecting PPO activity, which may be useful for protecting sprouts against PPO-related undesirable changes in their quality.

## 2. Materials and Methods

### 2.1. Chemicals

Catechol, Diethylaminoethyl–Sepharose (DEAE–S), tris(hydroxymethyl)aminomethane (TRIS), ethylenediaminetetraacetic acid sodium salt (EDTA), 4-methylcatechol, gallic acid, caffeic acid, L-cysteine, ascorbic acid, and DL-dithiothreitol were obtained from Sigma-Aldrich (Poznań, Poland). All other chemicals were of analytical grade.

### 2.2. Materials and Sprouting Conditions

Seeds from the lentil cultivar Tina were purchased from PNOS S.A. Ozarów Mazowiecki (Poland). The seeds were sterilized in 10% (*v/v*) sodium hypochloride for 10 min, drained, and washed with distilled water until they reached a neutral pH. They were placed in distilled water and soaked for 6 h at 25 °C. The seeds were dark-germinated for 4 days in a growth chamber on Petri dishes (diameter, 125 mm) lined with absorbent paper. Seedlings were watered daily with 5 mL of Milli-Q water [19].

### 2.3. Enzyme Assay

Polyphenol oxidase (PPO) activity was determined by measuring the initial rate of quinone formation, as indicated by an increase in the absorbance units (AUs) at 420 nm. An increase in absorbance of 0.001 $min^{-1}$ was taken as one unit of enzyme activity [20]. The increase in absorbance was linear with time for the first 120 s. The sample contained 1 mL of a 0.05-M substrate solution prepared in TRIS-HCl buffer (50 mM, pH 6.5) and 0.05 mL of an enzyme solution. All measurements were performed in triplicate.

### 2.4. Protein Determination

Protein content was determined according to the dye-binding method proposed by Bradford [21] using bovine serum albumin as a standard.

### 2.5. Enzyme Extraction and Partial Purification

One-hundred grams of sprouts were homogenized in 250 mL of 50 mM TRIS-HCl buffer (pH 6.5) containing 10 mM of ascorbic acid and 0.5% polyvinylpyrrolidone and were extracted with the aid of a magnetic stirrer for 1 h at 4 °C. The crude extract samples were centrifuged at 9000× *g* for 20 min at 4 °C. Solid $(NH_4)_2SO_4$ was added to the supernatant to obtain 80% saturation. After that, the precipitated proteins were separated by centrifugation at 9000× *g* for 30 min at 4 °C. The precipitate was dissolved in 60 mL of 5-mM TRIS-HCl (pH 7.0) and was dialyzed for 48 h using the same buffer in a cellulose bag with a membrane MWCO bigger than 12,000 Da at 4 °C. Afterwards, the dialysate was transferred to a DEAE–Sepharose column (20 × 250 mm) equilibrated with 5 mM of TRIS-HCl buffer, pH 6.5. Proteins were eluted, employing a linear gradient of 0 to 1.0 M of NaCl in 5 mM of TRIS-HCl buffer (pH 6.5) at a 30-mL·h$^{-1}$ flow rate. Three-milliliter fractions were collected, for which protein content (280 nm) and PPO activity toward catechol as a substrate were monitored. Fractions that showed PPO activity were collected.

### 2.6. Characterization of PPO

#### 2.6.1. Kinetic Data Analysis and Substrate Specificity

The specificity of PPOs from the lentil sprout extract was investigated for five commercial grade substrates (catechol, 4-methylcatechol, gallic acid, caffeic acid, and (+)-catechin) at concentrations of 1, 5, 10, 20, and 30 mM. The Michaelis constant ($Km$), maximum reaction velocities ($V_{max}$), and specificity ($V_{max}/Km$) of the PPOs were determined with the Lineweaver–Burk method.

#### 2.6.2. Effect of Temperature on Enzyme Activity

PPO activity was determined as a function of temperature in standard conditions at a temperature range of 20–80 °C. The optimum temperature for the PPO was determined using 50 mM of catechol as a substrate. The substrate solution was heated to the tested temperature, and then the enzyme solution was added. PPO activity was calculated in the form of percent residual PPO activity at the optimum temperature.

#### 2.6.3. Effect of pH on Enzyme Activity

PPO activity was determined as a function of pH in standard conditions using various buffers in the pH buffering range of 3.5–8.0 (3.5–5.5 acetate buffer, 100 mM; 5.5–8.0 potassium phosphate buffer, 100 mM). The optimum pH for the PPO was determined using 0.05 M of catechol as a substrate. The pH value corresponding to the highest enzyme activity was taken as the optimal pH. PPO activity was calculated in the form of residual PPO activity at the optimum pH.

#### 2.6.4. Effect of Antibrowning Agents on PPOs

The effects of ascorbic acid, citric acid, EDTA, L-cysteine, sodium azide, dithiothreitol, and sodium metabisulfite on PPO activity were examined. Three different concentrations of these inhibitors (0.2, 2, and 20 mM) were tested using 50 mM of the catechol substrate and were compared to a control enzyme reaction performed in optimal conditions with no inhibitor added. Percentage inhibition was calculated using the following equation:

$$\text{Inhibition } (\%) = (A_0 - A_i/A_0) \times 100\%, \quad (1)$$

where $A_0$ is initial PPO activity (without the inhibitor), and $A_i$ is PPO activity with the inhibitor.

#### 2.6.5. Effect of Ions on Enzyme Activity

The effect of ions, including $Na^+$, $K^+$, $Mg^{2+}$, $Zn^{2+}$, $Ba^{2+}$, $Fe^{3+}$, and $Mn^{2+}$ (chloride salts), on PPO activity was determined. Two different concentrations of these cations (2 and 10 mM) were tested using 50 mM of the catechol substrate. The effect of the studied ions on PPO activity was calculated in the form of percent residual PPO activity in comparison to the nontreated enzyme preparation.

### 2.7. Statistical Analysis

All data are presented as means including standard deviations (SDs) of three assays (means ± SD, $n = 3$).

## 3. Results and Discussion

### 3.1. PPO Isolation and Partial Purification

PPO was partially purified using a combination of ammonium sulfate precipitation and ion exchange chromatography (Figure 1). Two isoenzymes of PPO were found: PPO I and PPO II. The results of the purification of PPO are given in Table 1. After ammonium sulfate precipitation, the yield and purification fold were 90.6% and 4.67, respectively. The purification folds after ion exchange chromatography were 26.1 and 25.11 for the first and second isoenzymes, respectively. Further biochemical studies were performed on the first and second isoenzymes (important data in the enzymology field) and the total (crude extract) PPOs (data for food technology).

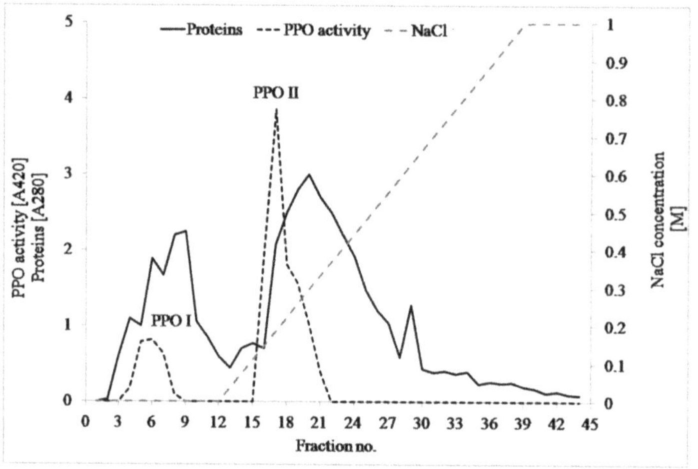


**Figure 1.** Anion exchange chromatographic elution profiles obtained after applying dissolved and desalted saline precipitate extract of lentil sprouts.

**Table 1.** Purification chart of polyphenol oxidases (PPOs) from lentil sprouts.

| | | Total Volume (mL) | Activity (U/mL) | Protein (mg/mL) | Total Activity (U) | Specific Activity (U/mg) | Yield (%) | Purification Fold |
|---|---|---|---|---|---|---|---|---|
| Crude extract | | 250 | 550 | 732.00 | 137500 | 0.75 | 100.0 | 1.00 |
| Salting out and dialysis | | 55 | 2265 | 644.97 | 124575 | 3.51 | 90.60 | 4.67 |
| Ion exchange chromatography | PPO I | 13 | 1815 | 92.56 | 23595 | 19.61 | 17.16 | 26.10 |
| | PPO II | 19 | 4475 | 237.12 | 85016 | 18.87 | 61.83 | 25.11 |

Catechol was used as a substrate for measuring PPO activity. An increase in absorbance of 0.001 $min^{-1}$ was taken as one unit of enzyme activity.

### 3.2. Substrate Specificity and Some Kinetic Parameters of Lentil Sprout PPOs

PPO kinetics were studied with four substrates, those commonly used for PPO assays (catechol, 4-methylcatechol) as well as those that are important from the nutraceutical point of view (gallic and caffeic acids). The $K_m$ and $V_{max}$ values calculated from the Lineweaver–Burk graphs are shown in Table 2. The values of $V_{max}$ and catalytic efficiency ($V_{max}/K_m$) indicated that 4-methylcatechol was the most suitable phenolic substrate for lentil sprout PPOs (Table 2). The $V_{max}$ values of total PPOs as well as PPO I and PPO II isoenzymes against gallic acid were also very high, but the $K_m$ of total PPOs was nearly twice and three times higher than the first and second isoenzymes. Most importantly, only PPO I used caffeic acid as a substrate ($K_m$ = 3.8 mM, $V_{max}$ = 769 U·mL$^{-1}$·min$^{-1}$). Total PPOs as well as PPO I and PPO II had the greatest affinity for catechol ($K_m$ = 1.32, 1.76, and 0.94 mM, respectively). These values were lower than those previously determined for persimmon ($K_m$ = 25 mM; sodium acetate buffer, pH 5.5) [15], green beans ($K_m$ = 10.6 and 37 mM for PPOI and PPOII, respectively; phosphate buffer, pH 7.0) [22], and mango fruit ($K_m$ = 10.6 mM, sodium acetate buffer, pH 5.6) [23]. All the studied PPOs of lentil sprouts exhibited the highest affinity for catechol: Hence, it was used as a substrate in further biochemical assays.

**Table 2.** Kinetic parameters of PPOs from lentil sprouts assessed with the use of several phenol substrates.

| | | $V_{max}$ (U·mL$^{-1}$·min$^{-1}$) | $Km$ (mM) | $V_{max}/Km$ (U·mL$^{-1}$·min$^{-1}$·mM$^{-1}$) |
|---|---|---|---|---|
| 4-methylcatechol | PPO I | 4878 ± 244 | 3.00 ± 0.14 | 1626 |
| | PPO II | 3846 ± 192 | 3.40 ± 0.15 | 1131 |
| | Total | 5410 ± 270 | 1.50 ± 0.07 | 3607 |
| Catechol | PPO I | 952 ± 48 | 1.76 ± 0.08 | 541 |
| | PPO II | 1111 ± 56 | 0.94 ± 0.04 | 1176 |
| | Total | 1737 ± 87 | 1.32 ± 0.06 | 1320 |
| Gallic acid | PPO I | 2817 ± 141 | 2.25 ± 0.10 | 1250 |
| | PPO II | 3742 ± 152 | 5.00 ± 0.23 | 769 |
| | Total | 8250 ± 413 | 7.25 ± 0.33 | 1138 |
| Caffeic acid | PPO I | 769±38 | 3.81 ± 0.17 | 202 |
| | PPO II | 0 | 0 | 0 |
| | Total | 0 | 0 | 0 |

All values represent the means of triplicate measurements.

### 3.3. *Effect of Temperature and pH on PPO Activity*

Figure 2A shows the influence of temperature on PPO activities in the assay conditions (pH 5.5 and 50 mM catechol as a substrate). PPO I, PPO II, and total PPOs reached maximum activity at 35 °C. The optimal temperatures for PPO activity are substrate-dependent and may differ for PPOs obtained from various sources [24]. It has been reported that when catechol is used as a substrate, the optimum temperature is 40 °C for PPOs from Chinese cabbage [25], soybean sprouts [26], and parsley [24]; and 25–30 °C for bananas [27]. Higher optimal temperatures were reported by Serradell et al. [28] and Navarro et al. [15] for PPOs isolated from strawberries (50 °C) and persimmons (55 °C).

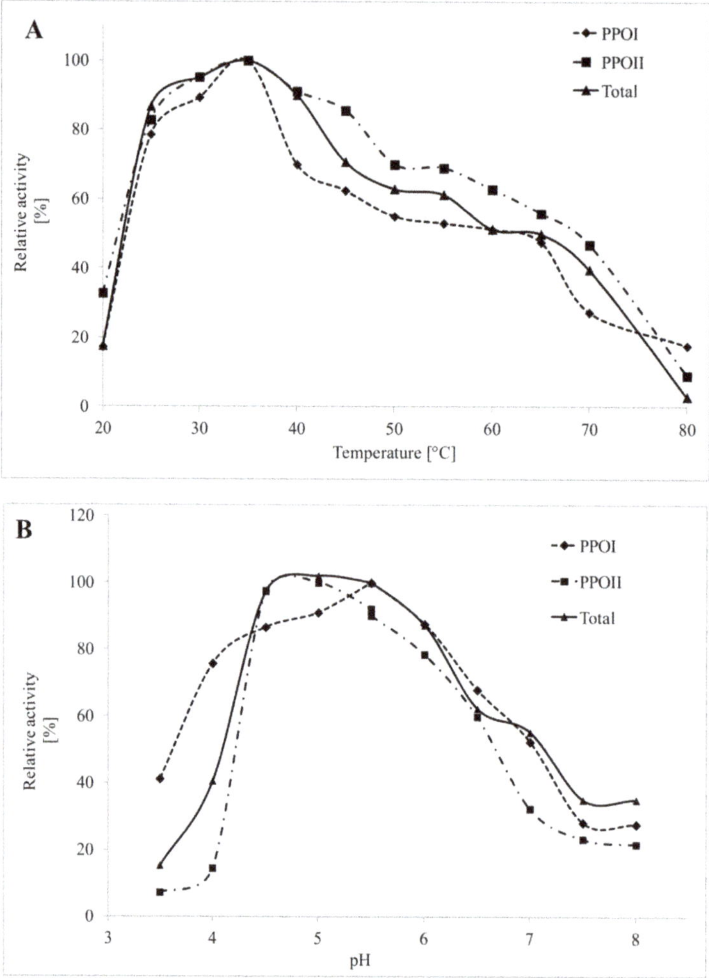


**Figure 2.** Effects of temperature (**A**) and pH (**B**) on the activity of lentil sprout PPOs.

The assay of PPO activity in a pH range from 4.0 to 8.0, using catechol as a substrate, showed optimal conditions for total PPO activity in the pH range of 4.5–5.5. When the two PPO isoforms were studied separately, a pH of 4.5–5.0 and 5.5 were found to be optimal for PPO I and PPO II, respectively (Figure 2B). The pH optimum for PPO activity has been found to be dependent on the enzyme source and purity, substrate, and buffer system used [1]. In the available literature, different pH optima for PPO activity have been reported. Two pH optima, suggesting the presence of two PPO isoenzymes, have been previously reported by other researchers, i.e., 4.5–5.0 and 7.5–7.6, for two PPO isoenzymes from avocados [29]; 4.0 and 7.0 for green bean PPOs [22]; and 5.0 and 7.5 for Jonagored apple PPOs [30]. A pH value of 5.5, i.e., the optimum pH for PPO activity determined in this study, agrees well with values that have been reported for lettuce [18] and mango fruit [23]. Contrarily, in a study performed by Nagai and Suzuki [26], PPOs from soybean sprouts exhibited optimal activity at pH 8–9.

### 3.4. Effect of Various Inhibitors and Metal Ions on PPO Activity

The effects of antibrowning agents on the activity of PPOs from lentil sprouts were studied using catechol as a substrate. The results, i.e., the percentage of inhibition relative to the control, are reported in Table 3. Ascorbic acid was the most effective inhibitor of total PPOs and showed ca. 62%, 43%, and 24% of inhibition of dark pigment formation at 20-mM, 2-mM, and 0.2-mM concentrations, respectively. The production of dark quinones by both isoenzymes was most effectively inhibited by 20 mM of ascorbic acid, L-cysteine, and sodium metabisulfite. These compounds also showed high efficiency at the lowest concentrations studied. Similar profiles have been previously found for PPOs from bananas [31], parsley [24], green beans [22], mango fruit [23], and soybean sprouts [26]. PPO II was also very sensitive to citric acid: Approximately 60% inhibition was recorded in the presence of 20 mM of the inhibitor. Furthermore, the purified PPOs were much more sensitive to the studied inhibitors than the crude extract, probably because some components were able to mask the inhibitory effect of these compounds. The degree of dark pigment formation depends on the origin of the PPO and substrate used, and thus it is difficult to compare the present results to other studies. Yagar and Sagiroglu [32] recorded 98% and 100% of inhibition of quince PPOs in the presence of 2 and 20 mM of ascorbic acid, respectively. For 2 and 20 mM of sodium metabisulfite, the degree of inhibition was 52% and 98%, respectively. As in other reports of the effect of antibrowning agents [11,20,33,34], the activity of lentil sprout PPOs was also inhibited by a thiol-containing compound (dithiothreitol) and a copper-binding ligand (sodium azide, EDTA), but these compounds are very toxic and are banned as food additives. Thus, of the studied antibrowning agents, only ascorbic acid, citric acid, and L-cysteine are suitable to be used in food technology.

**Table 3.** Effects of various antibrowning agents on the activity of lentil sprout PPOs.

| | Concentration of Compounds (mM) | % of Inhibition | | |
|---|---|---|---|---|
| | | PPO I | PPO II | Total |
| Ascorbic acid | 20 | 79.66 ± 3.03 | 79.42 ± 1.80 | 62.57 ± 2.38 |
| | 2 | 50.85 ± 1.93 | 59.42 ± 2.58 | 43.86 ± 1.67 |
| | 0.2 | 46.61 ± 1.77 | 62.32 ± 0.71 | 24.56 ± 0.93 |
| L-cysteine | 20 | 71.67 ± 2.72 | 72.09 ± 3.14 | 26.97 ± 1.02 |
| | 2 | 68.33 ± 2.60 | 55.81 ± 2.43 | 21.35 ± 0.81 |
| | 0.2 | 56.67 ± 2.15 | 34.88 ± 1.52 | 10.11 ± 0.38 |
| $Na_2S_2O_5$ | 20 | 76.03 ± 2.89 | 71.58 ± 3.11 | 25.88 ± 0.98 |
| | 2 | 56.20 ± 2.14 | 64.21 ± 2.79 | 14.12 ± 0.54 |
| | 0.2 | 18.18 ± 0.69 | 24.21 ± 1.05 | 1.18 ± 0.04 |
| EDTA | 20 | 26.67 ± 1.01 | 39.77 ± 1.73 | 14.63 ± 0.56 |
| | 2 | 24.17 ± 0.92 | 36.36 ± 1.58 | 3.66 ± 0.14 |
| | 0.2 | 20.83 ± 0.79 | 34.09 ± 1.48 | 2.44 ± 0.09 |
| Citric acid | 20 | 30.13 ± 1.14 | 60.82 ± 2.65 | 43.18 ± 1.64 |
| | 2 | 22.44 ± 0.85 | 54.39 ± 2.37 | 37.88 ± 1.44 |
| | 0.2 | 7.69 ± 0.29 | 57.89 ± 2.52 | 8.33 ± 0.32 |
| Sodium azide | 20 | 22.50 ± 0.85 | 22.89 ± 1.00 | 9.09 ± 0.35 |
| | 2 | 17.50 ± 0.67 | 15.66 ± 0.68 | 8.08 ± 0.31 |
| | 0.2 | 5.83 ± 0.22 | 8.43 ± 0.37 | 2.02 ± 0.08 |
| Dithiothreitol | 20 | 17.50 ± 0.67 | 37.89 ± 1.65 | 30.26 ± 1.15 |
| | 2 | 22.50 ± 0.86 | 31.58 ± 1.37 | 13.16 ± 0.50 |
| | 0.2 | 22.50 ± 0.86 | 15.58 ± 1.72 | 1.32 ± 0.05 |

All values represent the means of triplicate measurements.

The effect of metal ions on the activity of PPOs is presented in Table 4. $Zn^{2+}$, $Ba^{2+}$, $Fe^{3+}$, and $Mn^{2+}$ at a 10-mM concentration completely inhibited the activity of PPOs. In contrast, in studies conducted by Liu et al. [35], both $Zn^{2+}$ and $Mg^{2+}$ (10 mM) increased the activity of PPOs from flower buds of *Lonicera japonica* by about 10%–15%. In addition, Aydemir has reported [36] that the activity of

PPOs from artichokes was only slightly inhibited by $Zn^{2+}$ and $Mg^{2+}$ at 1- and 10-mM concentrations ($Fe^{3+}$ did not affect activity). Total PPOs, PPO I, and PPO II were inhibited by 2 mM of $MgCl_2$: 50%, 52%, and 67% of residual activity was detected, respectively. According to literature data, the effect of $Mg^{2+}$ on the activity of PPOs differs significantly and is strongly determined by the origin of the PPOs. The activity of PPOs from flower buds of *Lonicera japonica* has been activated by ions at 1–100 mM concentrations [35], whereas reduced activity has been noted at lower concentrations of ions (0.1–0.01 mM). An opposite relationship has been observed for PPOs from green beans [22] and artichokes [36]. There was no effect of 10 mM of $K^+$ on the activity of PPOs (except PPO I): However, at the 2-mM concentration, these ions activated PPO I, PPO II, and total PPOs. A similar pattern of relationships was recorded for total PPOs, PPO I, and $Na^+$ ions. It has been previously reported that $Na^+$ ions either did not affect or only slightly modified the activity of PPOs, e.g., from green beans [22], flower buds of *Lonicera japonica* [35], or artichokes [36]. On the other hand, the activities of PPOs from Ataulfo mango and Anamur banana have been inhibited by $Na^+$ ions [18,21].

**Table 4.** Effects of metal ions on the activity of PPOs from lentil sprouts.

| Ion Concentration (mM) | | Residual Activity (%) | |
|---|---|---|---|
| | | 10 | 2 |
| $Na^+$ | PPO I | 75.82 ± 3.26 | 82.42 ± 3.54 |
| | PPO II | 90.38 ± 3.89 | 142.31 ± 6.12 |
| | Total | 89.81 ± 3.86 | 106.48 ± 4.58 |
| $K^+$ | PPO I | 76.99 ± 3.31 | 118.58 ± 5.10 |
| | PPO II | 101.35 ± 4.36 | 159.46 ± 6.86 |
| | Total | 99.23 ± 4.27 | 109.46 ± 4.71 |
| $Mg^{2+}$ | PPO I | 46.90 ± 2.02 | 50.44 ± 2.17 |
| | PPO II | 51.35 ± 2.21 | 52.70 ± 2.27 |
| | Total | 61.54 ± 2.65 | 67.69 ± 2.91 |
| $Zn^{2+}$ | PPO I | Nd | 42.24 ± 1.82 |
| | PPO II | Nd | 38.42 ± 1.65 |
| | Total | Nd | 48.78 ± 2.10 |
| $Ba^{2+}$ | PPO I | Nd | 54.23 ± 2.33 |
| | PPO II | Nd | 54.43 ± 2.34 |
| | Total | Nd | 86.05 ± 3.70 |
| $Fe^{3+}$ | PPO I | Nd | 53.49 ± 2.30 |
| | PPO II | Nd | 56.96 ± 2.45 |
| | Total | Nd | 58.28 ± 2.51 |
| $Mn^{2+}$ | PPO I | Nd | 48.98 ± 2.11 |
| | PPO II | Nd | 32.92 ± 1.42 |
| | Total | Nd | 27.30 ± 1.17 |

All values represent the means of triplicate measurements. Nd: not detected.

## 4. Conclusions

Lentil sprouts are widely consumed all over the world. Enzymatic browning of sprouts during storage is a serious problem negatively influencing their consumer quality. Identifying and understanding the mechanism of inhibition of polyphenol oxidases (PPOs) in lentil sprouts may offer inexpensive alternatives in preventing browning. Our findings indicated that supplementation of sprouts with metal ions ($Zn^{2+}$, $Mn^{2+}$, $Fe^{3+}$) and/or inhibitors (ascorbic acid, citric acid) may be used for decreasing the activity of PPOs. This strategy seems to be justified, but more research is needed to define effects on the growth and metabolism of sprouts, as well as their nutritional and pro-health qualities.

**Author Contributions:** Conceptualization, M.S. and M.Ś.; methodology, M.S. and U.Z.; software, M.F.; validation, M.Ś., A.J., and M.F.; formal analysis, M.Ś.; investigation, M.S., A.J., M.F., J.B., and U.Z.; resources, M.Ś.; data curation, M.S.; writing—original draft preparation, M.S., J.B., and M.Ś.; writing—review and editing, M.S. and M.Ś.; visualization, M.S.; supervision, M.Ś.; project administration, M.S.; funding acquisition, M.Ś.

**Acknowledgments:** The authors would like to thank the Scientific Students Group of Food Biochemistry and Nutrition, Department of Biochemistry and Food Chemistry, University of Life Sciences in Lublin, Poland, for their technical support.

**Conflicts of Interest:** The authors declare no conflicts of interest.

## References

1. Yoruk, R.; Marshall, M.R. Physicochemical properties and function of plant polyphenol oxidase: A review. *J. Food Biochem.* **2003**, *27*, 361–422. [CrossRef]
2. Mayer, A.M. Polyphenol oxidases in plants and fungi: Going places? A review. *Phytochemistry* **2006**, *67*, 2318–2331. [CrossRef]
3. Vaughn, K.C.; Duke, S.O. Function of polyphenol oxidase in higher plants. *Physiol. Plant.* **1984**, *60*, 106–112. [CrossRef]
4. Chabanet, A.; Goldberg, R.; Catesson, A.M.; Quinet-Szely, M.; Delaunay, A.M.; Faye, L. Characterization and Localization of a Phenoloxidase in Mung Bean Hypocotyl Cell Walls. *Plant Physiol.* **1994**, *106*, 1095–1102. [CrossRef]
5. McDougall, G.J.; Stewart, D.; Morrison, I.M. Cell-wall-bound oxidases from tobacco (*Nicotiana tabacum*) xylem participate in lignin formation. *Planta* **1994**, *194*, 9–14. [CrossRef]
6. Zhang, Q.; Liu, Y.; He, C.; Zhu, S. Postharvest Exogenous Application of Abscisic Acid Reduces Internal Browning in Pineapple. *J. Agric. Food Chem.* **2015**, *63*, 5313–5320. [CrossRef]
7. Sun, Y.; Zhang, W.; Zeng, T.; Nie, Q.; Zhang, F.; Zhu, L. Hydrogen sulfide inhibits enzymatic browning of fresh-cut lotus root slices by regulating phenolic metabolism. *Food Chem.* **2015**, *177*, 376–381. [CrossRef]
8. Yi, J.H.; Dong, X.L.; Zhu, Z.B. Effect of polyphenol oxidase (PPO) enzymatic inducing factors on non-enzymatic browning of apple polyphenols. *Mod. Food Sci. Technol.* **2015**, *31*, 119–127.
9. Sikora, M.; Świeca, M. Effect of ascorbic acid postharvest treatment on enzymatic browning, phenolics and antioxidant capacity of stored mung bean sprouts. *Food Chem.* **2018**, *239*, 1160–1166. [CrossRef]
10. Sullivan, M.L. Beyond brown: Polyphenol oxidases as enzymes of plant specialized metabolism. *Front. Plant Sci.* **2014**, *5*, 783. [CrossRef] [PubMed]
11. Swieca, M.; Sęczyk, L.; Gawlik-Dziki, U. Elicitation and precursor feeding as tools for the improvement of the phenolic content and antioxidant activity of lentil sprouts. *Food Chem.* **2014**, *161*, 288–295. [CrossRef]
12. Ali, H.M.; El-Gizawy, A.M.; El-Bassiouny, R.E.I.; Saleh, M.A. Browning inhibition mechanisms by cysteine, ascorbic acid and citric acid, and identifying PPO-catechol-cysteine reaction products. *J. Food Sci. Technol.* **2015**, *52*, 3651–3659. [CrossRef]
13. Arias, E.; González, J.; Peiró, J.M.; Oria, R.; Lopez-Buesa, P. Browning prevention by ascorbic acid and 4-hexylresorcinol: Different mechanisms of action on polyphenol oxidase in the presence and in the absence of substrates. *J. Food Sci.* **2007**, *72*, 464–470. [CrossRef]
14. Mishra, B.B.; Gautam, S.; Sharma, A. Purification and characterisation of polyphenol oxidase (PPO) from eggplant (*Solanum melongena*). *Food Chem.* **2012**, *134*, 1855–1861. [CrossRef]
15. Navarro, J.L.; Tárrega, A.; Sentandreu, M.A.; Sentandreu, E. Partial purification and characterization of polyphenol oxidase from persimmon. *Food Chem.* **2014**, *157*, 283–289. [CrossRef]
16. Gawlik-Dziki, U.; Szymanowska, U.; Baraniak, B. Characterization of polyphenol oxidase from broccoli (*Brassica oleracea* var. botrytis italica) florets. *Food Chem.* **2007**, *105*, 1047–1053. [CrossRef]
17. Aydemïr, T.; Akkanlı, G. Partial purification and characterisation of polyphenol oxidase from celery root (*Apium graveolens* L.) and the investigation of the effects on the enzyme activity of some inhibitors. *Int. J. Food Sci. Technol.* **2006**, *41*, 1090–1098. [CrossRef]
18. Gawlik-Dziki, U.; Złotek, U.; Świeca, M. Characterization of polyphenol oxidase from butter lettuce (*Lactuca sativa* var. capitata L.). *Food Chem.* **2008**, *107*, 129–135. [CrossRef]

19. Świeca, M.; Gawlik-Dziki, U.; Kowalczyk, D.; Złotek, U. Impact of germination time and type of illumination on the antioxidant compounds and antioxidant capacity of Lens culinaris sprouts. *Sci. Hortic. (Amsterdam).* **2012**, *140*, 87–95. [CrossRef]
20. Cho, Y.K.; Ahn, H.K. Purification and characterization of polyphenol oxidase from potato: II. Inhibition and catalytic mechanism. *J. Food Biochem.* **1999**, *23*, 593–605. [CrossRef]
21. Bradford, M.M. A rapid and sensitive method for the quantitation of microgram quantities of protein utilizing the principle of protein-dye binding. *Anal. Biochem.* **1976**, *72*, 248–254. [CrossRef]
22. Guo, L.; Ma, Y.; Shi, J.; Xue, S. The purification and characterisation of polyphenol oxidase from green bean (*Phaseolus vulgaris* L.). *Food Chem.* **2009**, *117*, 143–151. [CrossRef]
23. Cheema, S.; Sommerhalter, M. Characterization of polyphenol oxidase activity in Ataulfo mango. *Food Chem.* **2015**, *171*, 382–387. [CrossRef]
24. Doğru, Y.Z.; Erat, M. Investigation of some kinetic properties of polyphenol oxidase from parsley (Petroselinum crispum, Apiaceae). *Food Res. Int.* **2012**, *49*, 411–415. [CrossRef]
25. Nagai, T.; Suzuki, N. Partial purification of polyphenol oxidase from Chinese cabbage *Brassica rapa* L. *J. Agric. Food Chem.* **2001**, *49*, 3922–3926. [CrossRef]
26. Nagai, T.; Suzuki, N. Polyphenol oxidase from bean sprouts (*Glycine max* L.). *J. Food Sci.* **2003**, *68*, 16–20. [CrossRef]
27. Ünal, M.Ü. Properties of polyphenol oxidase from Anamur banana (*Musa cavendishii*). *Food Chem.* **2007**, *100*, 909–913. [CrossRef]
28. Serradell, M.D.L.A.; Rozenfeld, P.A.; Martínez, G.A.; Civello, P.M.; Chaves, A.R.; Añón, M.C. Polyphenoloxidase activity from strawberry fruit (Fragaria x ananassa, Duch., cv Selva): Characterisation and partial purification. *J. Sci. Food Agric.* **2000**, *80*, 1421–1427. [CrossRef]
29. Gómez-López, V.M. Some biochemical properties of polyphenol oxidase from two varieties of avocado. *Food Chem.* **2002**, *77*, 163–169. [CrossRef]
30. Rocha, A.M.C.; Morais, A.M.M. Characterization of polyphenoloxidase (PPO) extracted from 'Jonagored' apple. *Food Control* **2001**, *12*, 85–90. [CrossRef]
31. Lee, M.-K. Inhibitory effect of banana polyphenol oxidase during ripening of banana by onion extract and Maillard reaction products. *Food Chem.* **2007**, *102*, 146–149. [CrossRef]
32. Yağar, H.; Sağiroğlu, A. Partially purification and characterization of polyphenol oxidase of quince. *Turkish J. Chem.* **2002**, *26*, 97–103.
33. Złotek, U.; Gawlik-Dziki, U. Selected biochemical properties of polyphenol oxidase in butter lettuce leaves (*Lactuca sativa* L. var. capitata) elicited with dl-β-amino-*n*-butyric acid. *Food Chem.* **2015**, *168*, 423–429. [CrossRef] [PubMed]
34. Gandía-Herrero, F.; García-Carmona, F.; Escribano, J. Purification and characterization of a latent polyphenol oxidase from beet root (*Beta vulgaris* L.). *J. Agric. Food Chem.* **2004**, *52*, 609–615. [CrossRef] [PubMed]
35. Liu, N.; Liu, W.; Wang, D.; Zhou, Y.; Lin, X.; Wang, X.; Li, S. Purification and partial characterization of polyphenol oxidase from the flower buds of Lonicera japonica Thunb. *Food Chem.* **2013**, *138*, 478–483. [CrossRef] [PubMed]
36. Aydemir, T. Partial purification and characterization of polyphenol oxidase from artichoke (*Cynara scolymus* L.) heads. *Food Chem.* **2004**, *87*, 59–67. [CrossRef]

MDPI
St. Alban-Anlage 66
4052 Basel
Switzerland
Tel. +41 61 683 77 34
Fax +41 61 302 89 18
www.mdpi.com

*Foods* Editorial Office
E-mail: foods@mdpi.com
www.mdpi.com/journal/foods

www.ingramcontent.com/pod-product-compliance
Lightning Source LLC
LaVergne TN
LVHW070625170726
843515LV00004B/180

* 9 7 8 3 0 3 9 4 3 3 1 6 2 *